BLACK IS THE NEW GREEN

BLACK IS THE NEW GREEN

MARKETING TO AFFLUENT AFRICAN AMERICANS

LEONARD E. BURNETT, JR. AND
ANDREA HOFFMAN

First published in 2010 by
PALGRAVE MACMILLAN®
in the United States—a division of St. Martin's Press LLC,
175 Fifth Avenue, New York, NY 10010.

Where this book is distributed in the UK, Europe and the rest of the
world, this is by Palgrave Macmillan, a division of Macmillan Publishers
Limited, registered in England, company number 785998, of Houndmills,
Basingstoke, Hampshire RG21 6XS.

Palgrave Macmillan is the global academic imprint of the above companies
and has companies and representatives throughout the world.

Palgrave® and Macmillan® are registered trademarks in the United States,
the United Kingdom, Europe and other countries.

ISBN: 978–0–230–61684–4

Library of Congress Cataloging-in-Publication Data

Burnett, Leonard E., 1964–
    Black is the new green : marketing to affluent African Americans /
Leonard E. Burnett, Jr. and Andrea Hoffman.
        p. cm.
    Includes index.
    ISBN 978–0–230–61684–4
    1. African American consumers. 2. Affluent consumers—United States.
3. Consumer behavior—United States. 4. Target marketing—United States.
I. Hoffman, Andrea, 1965– II. Title.

HC110.C6B87 2010
658.8'04—dc22   2009042479

A catalogue record of the book is available from the British Library.

Design by Newgen Imaging Systems (P) Ltd., Chennai, India.

First edition: March 2010

10 9 8 7 6 5 4 3 2 1

Printed in the United States of America.

CONTENTS

ACKNOWLEDGMENTS

My motivation for writing this book was the desire to set the record straight about the power of the Affluent African American (AAA) audience and the opportunity it presents to marketers today. AAA's are one of the most economically diverse groups in the United States. Their interests range from hip-hop to Bach. And their assets vary from Timex to Rolex. In these tough economic times marketers must look at new ways of growing their businesses. The AAA market represents an audience that is underserved with money to spend! Readers will find this book enlightening as well as surprising.

I would like to thank:

Andrea, my co-author, who gave me the idea and total support;
Marsha, my wife, who gives me love;
My children, Lenny and Rani, who give me joy;
My parents, who give me inspiration;
My brothers, who give me encouragement;
Brett, my business partner, who gives me trust; and
God, who gives me strength.

Thank you to all the readers along the way who will absorb what we have taken the time to share, and pass it along to others.

—Len

The idea to write *Black Is the New Green* was sparked by several forces: first, my 10-plus years of passionate dedication to gathering insights and research on affluent ethnic consumers and understanding how a lack of research could cause misperceptions among marketers; second, my firsthand experience seeing luxury brands often overlook affluent African Americans as a viable target market; and third, working with *Uptown* magazine and being inspired by Len's career accomplishments, hard work, and dedication. But, truth be told, my original goal was to land Len a book deal so he could share his story about being an urban media pioneer. Then a funny thing happened.

I met with the brilliant Airie Stuart, vice president at Palgrave Macmillan, who sat and listened intensely to the book idea (and even canceled her next meeting to finish listening to the pitch). Somehow, by the end of the meeting, I was co-authoring a marketing book about affluent African Americans. Airie clearly has vision. It wasn't long after our meeting that President Obama was elected. Timing is everything. I thank Airie for "getting it" and providing Len and me with a platform from which to share our knowledge. Thank you for being a blessing and a visionary.

I would also like to thank a personal friend and colleague, a man I much admire on many levels—talented executive, dedicated humanitarian, giving soul, and munificent philanthropist—Reginald Van Lee of Booz & Company. It was Reggie who introduced me to Airie after I mentioned my idea. Thank you for being you. I aspire to possess just a few of the traits you innately have.

To our patient and talented editors Laurie Harting and Lynn Vannucci, who kept Len and me focused and clearheaded when the pressure was on. Thank you for lending us your skills for these months. I think we'll have withdrawal not working with you anymore!

Len and I are grateful to the following contributors to the book. Your support, encouragement, and participation were truly a gift. In today's challenging business environment, we all have very little time to give, but each of you made time for this important topic. Gregory Furman of the Luxury Marketing Council; Bob Shullman, president, Ipsos Mendelsohn; Dwayne Ashley, CEO of Thurgood Marshall College Fund; Byron Lewis, chairman, Uniworld Group; Marc

Bland, manager, Analytical Solutions/Multi-Cultural Marketing Lead at R.L. Polk & Co.; Tracy Ulrich at Harley Davidson; Adam Goldman of Human Applications; Carl Brooks, president and CEO, Executive Leadership Council; Soledad O'Brien; Lorenzo Benazzo, CEO of Hyphen; Constance White of eBay; Kadesha Boyer at Sony Electronics; Nancy Armand of HSBC; and Noel Hankin, senior vice president, Multicultural Relations, Moet Hennessy USA. A huge thank you to Christopher Vollmer of Booz & Company for giving generously of his time to be interviewed for the book, and Jim Clifton, CEO, The Gallup Organization, who made himself available for an endorsement.

Special thanks to Kojo Bentil, who has been a huge support system to me for the last year; Amy Bassi, the most talented project manager a business owner could wish for; Valerie Chaples for her friendship and sound business advice; Ivan Burwell for believing; Joyce Mullins Jackson for connecting; and Bernard Hampton Jackson III for listening. And of course, my parents Paul and Wilma Hoffman for remaining excited about seeing this book completed and pushing me, even when I was exhausted and out of ideas. They should write their own book about being married for over fifty years!

Last, but not least, Len Burnett for his respect and for making @#$% (things) happen!

—Andrea

BLACK IS THE NEW GREEN

Introduction: Where the Money Is

"Go where the money is." This pithy way to navigate should be obvious to any marketer worth his or her salt. It's known as "Sutton's Law," and it comes to us out of the Great Depression, by way of a colorful character known as "Slick Willie" Sutton. Sutton was a prolific bank robber known for his immaculate dress, quick wit, and gentle manners. Although the sheer number of heists he pulled off made bank thieves like John Dillinger look like amateurs, Sutton never engaged in violent behavior. Sutton is best remembered today for his reputed answer to a question from the reporter Mitch Ohnstad: Why do you rob banks?

"Because that's where the money is."

Though Sutton never actually uttered these words—he admits as much in his autobiography,[1] speculating that Ohnstad "invented" this answer, probably to fill out his story—the exchange has become urban legend. The sentiment has since been fashioned into an instrument for teaching medical students, forms a principle of activity-based costing (ABC) of management accounting where it is known as "Sutton's rule," and has become shorthand for simple common sense to anyone who has a product or service to sell.

At one time "go where the money is" may have sounded like a fairly straightforward directive. There was a time about forty or fifty years ago when it might have been true—a picture of a time we are admittedly painting with broad strokes. Though the population of the United States has always been richly diverse, founded as it was, by immigrants and proudly proclaiming itself a "melting pot," day-to-day life was lived out within a rather strict class system. There was an obvious shorthand for figuring out who fit where in the scheme of things: fur coats, big cars, and certain Anglo-Saxon surnames belonged to rich people; the suburbs were reserved for the middle class who, like the rich folks, were assumed to be White; working-class and lower-class neighborhoods were for the most part peopled by the latest waves of immigrants who— almost invariably—lived within their own ethnic clusters, laboring in the least desirable jobs, struggling to find their footing on the class ladder so that their children could make it up a rung or two. Aspirationally, if not culturally, the country *appeared to be* homogenized. The manner in which a centralized media portrayed and reported about American lives certainly lent itself to that conclusion, and targeting those who would use certain products and services within that media was a no-brainer.

But the days when plying the craft of marketing was limited to presenting a client with a clever layout for a general print campaign designed to appeal to the "typical" consumer are long gone, as is the "typical" consumer himself. The eulogy for that old advertising era would be the recent hit television show *Mad Men*. This eloquent look back is proof of how far we've moved on from the era—the early 1960s—the program showcases. Part of the show's appeal rests on nostalgia, and not necessarily the kind that women, African Americans, or most other minority groups revisit with pleasure.

So, if a return to that more culturally constrained and technologically innocent time isn't possible, let alone desirable, how do marketers find their footing today when the task of defining and reaching customers and potential customers is more complicated and downright difficult?

The first step is to recognize that business is entering a brand new era. It is fresh and, for the most part, unexplored territory. And, as it has been during every other time in history when change was in the air, those who stick to old ways of thinking and doing things—traditional practices and/or technologies that are fast becoming obsolete—miss the boat while those who adapt sail smoothly into the future. And the only way to keep your company and your brand from becoming irrelevant is to reach out and really figure out which way the wind is blowing. We've identified the four converging gale-force winds that are knocking around business—and specifically the luxury brand market. The purpose of this book is to show media, marketing executives, brand managers, business development experts, television programmers, Internet content developers, and others how to harness all that natural power and use it to their advantage.

FORCE #1: THE ECONOMY

The recessionary summer of 2009, when we were writing this book, is a long way from the heady days of the 1920s and the turbulent 1930s when Willie Sutton was active. For one thing, we are literally no longer so sure about where the money is. There is no longer any guarantee that it's in the banks or the investment firms, or any of the other places we are used to finding it. To say that this time of financial realignments has left marketers scrambling to find out who *still* has the money—and what would make them part with it and how often—is an understatement. In this era of shrinking consumer spending—and the correlating contraction of marketing budgets—there is increased pressure to squeeze each penny of worth out of every marketing dollar. How do you do that?

To begin with, by honestly and critically assessing current events to gain an understanding of what they mean for the long term. We don't have to look too far into the past to find a prime example of what happens when a company gets stuck in a time bubble. We have to look back only about ten or fifteen years and wonder where the American auto industry might be today had its leaders perceived the subtle signs of market change, where concerns about global oil supplies and rising gas prices created consumer demand for smaller, fuel-efficient cars and fuel efficiency inevitably impacted consumer buying patterns, and responded early enough to build automobiles that today's consumers are clamoring for.

What might have happened had the auto industry matched design innovation with marketing innovation that took advantage of technologies that, a decade or so ago, were just beginning to explode? What if auto executives had hired a trend forecaster and taken to heart what some in the market were clearly saying even way back then, that the "general market" was slowly but irreversibly going the way of the dinosaurs—just like those guys in *Mad Men*?

If the image of once-powerful car companies approaching the American taxpayer with hat in hand to stave off bankruptcy struck a chord with you, sending a chill of fear up your spine, it should. Those in the marketing world understand that to be nimble and foresighted is to be successful; anyone who is confronted with a clumsy failure of a magnitude the one the American car industry is facing might be fearful of how "spry" they really are.

FORCE #2: TECHNOLOGY

Certainly the market downturn that stunned us in 2008 is not the only force shaping the emerging new era. The art of marketing is itself in flux—or, if you prefer, on a learning curve—as advertisers, ad agencies, and even their media partners struggle to figure out how to fit the technologies profitably into the old mix of print, television, and radio formats and how they can create a new mix that will continue to deliver results for their clients. Think again of the American car companies and how their fate would have been different if, ten or fifteen years ago, they had the foresight to fashion marketing outreach programs that would have engaged consumers in an age when social media and other advanced communications technologies would be driving consumer demand.

Even today, while nearly three-quarters[2] of agency respondents to a recent survey said they expected to significantly shift spending to online advertising within the next two to three years, "only about one-quarter of marketers regard themselves as digitally savvy, and half claim they lack support at senior levels to substantially increase the marketing dollars allocated to digital media."[3]

To understand just how disoriented established institutions remain about how to deal with the new digital age, consider the plight of the nation's newspapers. *The New York Times* adapted early to the digital format, delivering the gold standard of reportage to over twenty million online viewers a month, *for free*. Theoretically, the subscription dollars lost by giving away free content would be made up with increased online advertising. As of this writing, that hasn't happened. Why?

The newspaper is still floundering financially because the only thing it did, essentially, was to transfer the content of its printed pages to a digital format for screen delivery. On the Internet, readers aren't confined to the information contained in just one paper's packaging of the news, and they aren't inclined to stick around any one particular site unless there's a compelling reason to do so. *The New York Times* hasn't yet found its digital niche and, importantly, it hasn't yet found a way to monetize its digital edition.

It would be unfair, however, to single out the *Times* in this regard. *The Wall Street Journal*, for example, entered the digital age enjoying a built-in niche as the country's leading instrument of financial and business news. But although the *WSJ*'s digital edition provides some free content, it has always been a subscription operation, and it isn't profitable yet either. Adding to the confusion is the contention by some leading digital thinkers, like Chris Anderson, editor-in-chief of *Wired* magazine and the author of the controversial book, *Free: The Future of a Radical Price*, that *free* is the new business model. "The marketplace wants free," Anderson says. "Consumers want free, and if you decide to set up a subscription service, then your competitor will make a free one."[4]

The bottom line is, if online advertising and/or subscription services are an ongoing experiment for such venerable institutions as the *New York Times* and the *Wall Street Journal*—as well as for such relatively recent media phenomena like YouTube, which is still losing money too, in spite of an audience size that rivals that of old-time TV stations—it's clear just how problematic adapting new technology to our purposes can be for the rest of us.

FORCE #3: SPENDING FOR MEANING

But, let's face it, technology is going to continue to evolve and consumers will be driven by advances in both products and methods of communication. In fact, science and the development of new technologies are progressing at such a rapid pace that a young person who entered a four-year college program in 2009 can expect that what he or she learned about science in freshman year will be outdated by the time he or she is a junior.

This rapid pace of change notwithstanding, our use of technology will necessarily become more refined. Just as a hundred years ago businesses gave up coal-fired machinery and adapted to new electricity to produce their goods more efficiently and cost-effectively, our understanding of how to employ the Internet, social networks, text messaging, Twitter, and whatever comes next in our marketing efforts will become more efficient and cost-effective as well.

As for the economy, downturns are cyclical; we'll likely remember the financial crisis of 2008–2009 as having required some extraordinary measures to affect correction, but it will indeed, at some point, become a memory.

What we can expect to stay with us long term as a result of the financial correction is a profound shift in customer spending—the end of the era of *irrational retail consumption*.

What exactly does that mean?

In recent decades, fueled by a seemingly infinite capacity to generate riches, consumers went on a binge. We bought and bought and bought some more, and when our houses became too small to hold all of our things, we built bigger houses. Consider this: in 1970, the average family of four lived in a house that was 1,400 square feet; in 2004, the average family of four lived in a house that was 2,330 square feet.[5] In over thirty years, the size of our living space has almost doubled—and we've filled all that extra space with all the furniture and rugs and art and clothes and toys we've collected during our decades-long spending spree.

Because we were so good at buying things to fill up our big houses, retailers got spoiled, expecting the orgy would never end. Within this same time frame we even found a whole new way to shop: the Internet. Sites such as eBay provided us with an international outlet to shop for anything we wanted anywhere in the world and, in the case of the African American audience, Internet shopping leveled the "race" playing field as it existed in traditional brick and mortar stores where race still plays a role in how the consumer is treated.

But the truth of it is that the sheer volume of the spending was not natural and could never be sustained. Now, though the financial crisis of 2008–2009 was certainly a catalyst for the consumer to reevaluate his or her relationship with money—and for marketers to reevaluate their relationships with their advertising budgets—we aren't experiencing a recession-fueled drop off in sales as much as we are moving, en masse, to embrace more realistic spending habits.

Thrift is no longer a dirty word as consumers are rethinking the meaning of value—though we hasten to add that even the interpretation of the word "thrift" may be in flux. In times of recession, the media creates a one-sided vantage point that people aren't spending at all when, in fact, what is happening is far more subtle: people *are* spending but they are doing it more wisely, and assessing the outflow in a more holistic way, in terms of how the dollars that are expended enhance the quality of their total life experience. *Experiential* is the new buzzword as buyers shift to placing more importance on the cultural activities involved in a vacation rather than in its R&R potential, the camaraderie of an evening of fine dining rather than merely the food and wine, the *experience* of a theater or concert event over the *acquisition* of material goods. We believe that this emerging new attitude about values and what makes something worthwhile will be a lasting one. It's a sea change in consumer behavior and leadership-driven marketers are going to find ways to position their brand to become a relevant part of their customers' lives.

FORCE #4: DEMOGRAPHIC SHIFT

Adapting to the new economic realities is compulsory for any business that wants to remain viable, but it doesn't mean making a huge stretch. It does not mean a vast and expensive restructuring of marketing plans or product lines. Spending time and money to look far afield is neither efficient nor necessary when the answers are often right under your nose. Your organization may be small or it may be complex, but the solutions to your new marketing challenges can often be quite simple, a matter of analyzing what equities and assets you may already be invested in and finding out that they can be used to gain a whole new audience (what opportunities are already knocking on your door or waiting right outside of it) and leveraging them in new ways. The place to *begin* is with a clear understanding of what the new market realities are. And that understanding is embedded in a firm grasp of how demographics we once took for granted are shifting seismically.

Let's start with a semantic shift. The word "minority" can mean a group of people that is lesser in number in comparison to the "majority." The number of people who own a Lexus is less than the number of people who own a Honda Accord; in Congress, in 2009, the Republican party is in the minority when compared to the Democrats; for a long time, Blacks, Hispanics, Asians, and members of other ethnic groups were a minority in comparison to the population of Whites. But the term "minority" as a label for these ethnic groups is fast becoming obsolete. In fact, by the year 2042, Blacks, Hispanics, Asians, Native Americans, and Pacific Islanders collectively will outnumber Whites. Among children under eighteen years old, they will outnumber Whites by the year 2023.[6] Indeed, in the years between 2000 and 2007, the ethnic population in the United States grew at rates almost three times that of the total population, and it now stands at 31.6 percent of the total population, compared to just 28.6 percent in 2000.[7] The writer Hua Hsu frames the meaning of these statistics in a way that cuts right to the chase: "every child born in the United States from here on out will belong to the first post-white generation."[8]

Indeed, in some places the shift has already happened. The nation's largest states, Texas and California, already have 'minority' majorities.[9] It's a trend forecast as long ago as 1998 by someone as ultimately, and intimately, tapped in as Bill Clinton: "In a little more than fifty years, there will be no majority race in the United States. No other nation in history has gone through demographic change of this magnitude in so short a time..."[10]

And just like Whites, ethnic consumers come in all shades and socioeconomic backgrounds. So the growing affluence within these former minority groups is of even more importance to the luxury marketer. The total number of affluent ethnic households in the United States in now estimated at over 1.3 million. The total income of these households is estimated at $387 billion with $150 billion in the hands of affluent Asian American households, $121 billion in Hispanic households, and $107 billion in African American households.[11]

Let's hone in on what this means within just one of these particular groups: the buying power of affluent African Americans (we will refer to them in this book as "AAAs") is currently $87.3 billion.[12] "Buying power" means, of course, "disposable income," the total after-tax income available to any individual to spend discretionally on whatever it is he or she likes or wants to have. Eighty-three *billion* dollars: a number that's guaranteed to increase with the new census study.

Overall Black buying power is expected to reach more than $1.1 *trillion* by 2012—just two short years for a cumulative growth of 28.4 per cent.[13]

It would be foolish in the extreme not to tap into this rich buying segment, yet that is exactly what the marketing arms of companies do all too frequently. Sometimes this is because the executives in a particular marketing department are unaware of the potential that exists within this segment. Sometimes it's because they are baffled about how to reach out to this segment. Sometimes it's because they think they lack the money or resources to make a credible effort at adding a whole new segment—especially when, as in today's economy, their marketing budgets are likely already stretched. And sometimes, unfortunately, it's because they *have* reached out in the past but, because their efforts were unappealing to the AAA audience and/or unsustained, they've not seen the return they ought to see for their investment dollars.

This is where we can help. Affluent African Americans are a segment we happen to know a great deal about. Why? We've made it our business to know. We've done the research because this is where the money is.

Who are "we"?

We are Andrea Hoffman and Leonard E. Burnett, Jr. We have a combined total of over forty years catering to this market segment.

For over twenty years, Andrea has been a marketing specialist and trend forecaster. She's worked with BMW of North America, Mercedes-Benz, Alliance Capital Management, National Council of La Raza, Sony Electronics, and 20th Century Fox Television. Among her accomplishments for these companies are the engineering of high-profile marketing strategies and events with well-known athletes and other celebrities. She was the key architect of the Mercedes-Benz USA's (MBUSA) first formal and highly successful entertainment strategy that included several diversity tactics, and she also was part of a team that helped to evaluate MBUSA's multimillion dollar Presence Marketing budget during a time when Mercedes was shifting its brand toward a younger demographic. She is a specialist in business development, diversity initiatives and marketing audits. She is the founder and current CEO of Diversity Affluence (www.diversityaffluence.com), a research, marketing communications and business development consultancy that helps brand marketers, agencies, the media, and entrepreneurs to understand and market to affluent ethnic consumers—a group for which Andrea has trademarked the term "Royaltons."

Len is a pioneer in urban media, and has been transforming marketers' perceptions of the importance of reaching the underserved urban audience for twenty years. He is one of the original team members of *VIBE*, the nation's leading urban music and culture magazine, and of its spin-off, *VIBE Vixen*, the fashion and beauty title for women. In this venture he collaborated with his lifelong friend and business associate Keith Clinkscales, the legendary Quincy Jones, and Time, Inc. As the magazine *Fast Company* put it in an interview with Len in 2007, he and Clinkscales "cut a wide swath helping to create a market for urban, selling to print media who had never considered the market, giving record labels a far reaching platform to promote hip hop music (roughly 800,000 readers), and giving fans of all colors an important touchstone for the culture."

Len went on to co-found Vanguarde Media, where he was Group Publisher of the magazines *Honey*, *Heart & Soul*, *Impact*, and *Savoy*. He is currently the co-CEO and Group Publisher, along with Brett Wright, of Uptown Media Group, which publishes *Uptown* magazine, a bi-monthly publication for urbanites focusing on the lifestyle and culture of AAAs—and which also comprises UptownLife.net, the interactive Web site, as well as *Uptown*'s signature and advertiser sponsored upscale events, and Uptown Social, Uptown's answer to Daily Candy, a weekly e-mail newsletter reflecting the little luxuries and indulgences this audience appreciates.

Together and separately we have been working with the purveyors of luxury goods and services to successfully target the AAA market for decades. We know this audience inside and out.

What we don't know is why so many others haven't yet tapped into the same potential—or wouldn't recognize AAAs as prospective customers if they did. The facts and figures we've cited in this chapter about demographic shift and income growth are a matter of public record, and Hsu claims, "Tellingly, every person I spoke with from [the advertising, marketing, and communications] industries knew the Census Bureau's projections by heart."[14] We can only assume then that the neglect of this audience is, in large part, due to marketers not yet having a firm grasp on how to approach this evolving target segment or how to repurpose marketing dollars towards new and fruitful opportunities. So we have decided to share our research as well as our firsthand experience to help others with understanding this critical niche market, shorten the learning curve to reach them, and allay any fears of risk a marketer might have in targeting a whole new consumer group. Because niche and target marketing is no longer something that is "nice to do"; it's a "need to do" in today's marketplace. We can't represent the urgency in a more straightforward way than to tell you this: any marketer who wants to produce results and grow business for his or her clients simply will not be successful if they do not understand the "profound demographic tipping point" on which we are now poised, and do not act to make their brand relevant to the emerging new majority.[15]

So, how do we start you on your way toward seizing this golden opportunity that so many have been overlooking, to their peril? Well, for starters, profile a typical AAA:

- She's a Senior V.P. of marketing for a major TV network.
- She drives a luxury car.
- She recently built her home in an upscale neighborhood.
- She dines out often with friends at exclusive restaurants.
- She's in her 40's, single, and has an MBA.
- Her annual salary? $125,000.
- Oh, and by the way, she's African American.

The typical AAA is not urban. She is not hip-hop. He's not an athlete or any other kind of celebrity. She made her money the old-fashioned way: working her way up the corporate ladder, bolstered by a solid educational background. He represents what happened to the Huxtable kids, the sons and daughters of Black professionals who were introduced to the public at large by Bill Cosby in the 1980s. He is the antithesis of the stereotypes and misperceptions that still, almost ten years into the twenty-first century, stand in the way of effective marketing programs targeted at AAAs. Let's talk about some of those stereotypes and misperceptions about African Americans, however uncomfortable discussing our preconceived notions can often make us.

- If I have money, I made it as a sports or entertainment figure.
- Some of us may be "rich," but none of us are "wealthy."
- We are all social liberals.
- Our values are somehow at odds with traditional American values.
- Most of us have poor credit, spend beyond our means, or don't pay our bills.
- I probably grew up in a single-parent home.
- When we aren't speaking in "urban slang" we "sound White."

As writer Moses Foster put it in a May 2008 article for *Advertising Age* in which he spoke about how the influential (and disappointing) power of the stereotypes: "I don't talk white. I talk like I've got $100,000 of education invested in me."[16]

It is just these sorts of disappointingly lingering stereotypes that can lead a company to miss out on the opportunity to market to AAA professionals—or to believe, dismissively, that they don't have to put money against reaching this new and emerging market because they've got AAAs covered in a general marketing strategy. Don't quite get what we're saying yet? Ok, take a look at another prospect.

- He's president of a large, multinational enterprise.
- He has two cars: one with a driver for work, and another just for fun.
- He has the use of a jet for special occasions.
- He entertains an international list of dignitaries.
- He has a large house with a staff in one city, and a smaller one nearer to family.
- He's an avid golfer and a student of history.
- He's 46, married to an executive, and they have two kids.
- Their combined annual income? Around $300,000.

Well, that was before he became the 44th president of the United States.

If you didn't recognize Barack Obama's resume—or if you did but dismissed the accomplishments listed there as out of the ordinary for an African American—you are surely missing the boat as it sails off into the future.

And, make no mistake about it, there are going to be people who miss the boat. At a party last spring, we were catching up with a few friends we hadn't seen for a while. In the course of catching up, we mentioned that we were also in the process of writing this book, about how to market to the affluent African American segment. A man who was also at the party, who we hadn't yet met, overheard and tossed off the comment: "You're going to market to affluent African Americans? All ten of them?"

This was a comment based on perceptions and stereotypes related to race and, as is the case with all such comments, its speaker was deeply misinformed. It is countered by no less an authority than Gregory J. Furman, the founder and chairman of the Luxury Marketing Council. According to Furman:

> It's no secret that the luxury consumer no longer consists only of celebrities and people who've inherited wealth (only ten percent of the luxury buyer universe). The working middle class with middle class values who've made it on their own AND niche communities which include women, the Latino/Hispanics, Asians, gay and lesbians, and importantly today, the black community—who, up until recently, have not been aggressively, intelligently, and creatively courted by luxury brands. To paraphrase my dear mentor—Mr. Stanley Marcus, it's high time luxury brands saw the community of highly affluent black Americans as essential to their business growth especially in these challenging times. All it takes is the sensitivity and know-how.

So, if, like Greg Furman, you want your brand in the 343,300 African American households in the United States that enjoy a minimum income of $150,000 or more per year—and if you want the 819,700 individuals who earn a minimum of $75,000 per year as part of your customer base, then get on board. We'll help you acquire the sensitivity and know-how to design complete plans or pilot programs that will be effective. We track the spending habits and lifestyle choices of AAAs for a living. We market to them, we are friends with them, we go to fundraisers where they converge and we can show you how to reach out to this underserved group, and how they will become a new target audience for you. We can show you why they are not always easy to reach via the general market or mainstream marketing platforms. We can help you develop the "360-degree" marketing style that attracts and engages this vibrant community. We can show you how this outreach can be the difference between your brand merely surviving, or actually thriving. We can show you how increasing your company's bottom line through the creation of brand loyalty among this important demographic segment can enhance your career.

Now, we're sensitive to the fact that, over the years, there have been some seemingly conflicting messages about this segment—insights that are more often than not the result of convoluted interpretations or inexpert research, or simply a lack of research. These mixed messages have, frankly, been based in outdated and/or one-dimensional thinking ("Isn't this just another diversity thing?"), or poorly thought-out promotional ideas ("Those billboards in Harlem didn't work!"). The mixed messages—and mixed results of sub par marketing campaigns—are what results when a marketer doesn't truly know the audience he or she is trying to reach. You see, with this group, it's all about really marketing to them—getting their loyalty, making them advocates for your brand and creating a relationship that will extend as their own wealth increases. It's all about learning what their needs are, how they define luxury, their buying and shopping habits—even beyond luxury buying and shopping—and connecting to what luxury and value mean to them. It's about how to reach them, and *where* to reach them.

For this segment of the African American community, luxury has reached critical mass in two ways. First, the sheer number of AAAs is now large enough for them to represent a profitable marketing segment—and unbeatable prospect base. Second, their disposable income is large enough to support spending habits they no longer view as luxurious, but merely part of their everyday life. Consider the AAA who took part in one of the focus groups we convened as part of our research. When we presented a list of luxury car makers to her, asking which of these high-end brands she might own or aspire to own, she was nonplused that BMW had made our list. A BMW, she informed us, was not a luxury vehicle, it was simply the car she drove every day; a Rolls Royce, on the other hand, that was a luxury vehicle to aspire to. Not all, but many AAAs grew up in upwardly mobile households, the children of parents whose high expectations and investment in their child's education and social contacts created a familiarity with brands above the level of mass market—and, clearly, above even what in the mass market passes for luxury. Their parents were the first generation to be specifically targeted by marketers but the brands that reached out to this pioneer generation may no longer resonate with the aspirations of their children.

Even more to the point, this segment of the market has an innate understanding of the power and influence they now wield in the marketplace, as well as *confidence in the ways that they relate to their growing wealth*. Most of the members of this newest generation of consumers are in the "spending prime" of their lives: they are deeply into their professions and careers, they have established families, they have created or expanded upscale communities, they take luxury for granted, and they support their community's social and philanthropic organizations on a grand scale. Any brand that would seek to grace these environs will need to prove to AAAs that it *belongs* there.

Just as mass marketers once proved to their parents that their brand belonged, to achieve "share of mind"—or claim a share of the vast AAA disposable income—luxury marketers, as well as marketers in general, must know who they are and where they live and what speaks to their unique assessment of what constitutes value, and *values*. As those mass marketers are now reaping the long-term results of targeting this segment, investing in marketing to the AAA audience will reap rewards now and in the future. It's 2010. An AAA is in the White House. The political clout of this group is starting to match its economic clout. It's time to start the coming out party. In that spirit, we welcome you to the new niche in town: affluent African Americans.

The general market for luxury goods has stagnated. Given the new economic reality of the early twenty-first century—not to mention the all-important new demographics of the new century—it's imprudent to continue to rely on luxury's traditional customer base to support sales, or on tired marketing strategies and tactics. At least, it is for businesses that want their market share to grow. In this book we are going to show you how to follow in the footsteps laid down by brands such as Gucci, HSBC, Sony Electronics, Aston Martin and others to become successful in a segment you can't afford to overlook if growth is your objective.

If you want to go where the money is, turn the page.

CHAPTER 1

CORPORATE AWARENESS

Olatz sheets and an Isomac Alba espresso maker. An Hermes tote bag for her and a custom-tailored business suit for him. His Mercedes S500 for driving to work and to workouts with his personal trainer and her Mont Blanc pen for composing her personal correspondence. It may seem obvious why the typical affluent African American (AAA) chooses quality products. They are well-made and long-lasting items that the purchaser can take pleasure in. But how does an AAA decide on a particular luxury brand? Why does she own an Hermes tote as opposed to another brand? Did she simply prefer the topstitching on the Hermes bag? Of all makes of cars on the market, why did he choose to buy a Mercedes? And what of the myriad nonmaterial decisions an AAA makes, such as her hairdresser, his travel agent, and the family financial advisor?

These are questions that any marketer wants the answers to and there are very specific answers, indeed, because the typical AAA does not make random choices. We will explore these answers later, but before we do, we'd like to give you a framework in which to understand them. In this chapter, we will examine four major pitfalls in corporate thinking that can often obstruct a well-intentioned effort to market to the rising AAA consumer demographic.

To this end, let us first look at a personality test that was developed over sixty years ago. Called the Johari Window, this test has now been adapted by top management consulting firms to help corporations strategically plan their futures. The test in this adaptation is a very useful exercise for the marketing executive—a bit of yoga for the brain that will help you stretch and relax any stiff assumptions you may have about this market segment.

GETTING TO KNOW YOUR TARGET MARKET

In the 1950s, two doctors at UCLA, Joseph Luft, PhD, and Harrington Ingham, MD, were researching the human personality when they devised the now-famous Johari Window. This test doesn't measure personality; rather, as the name implies, it provides a "window" into how personality is expressed, a way of mapping personality awareness.

The Johari Window is a simple test, and actually a fun one, too. First, you are given a grid comprised of a fixed but extensive list of adjectives—words like "able," "dignified," "independent," "logical," "loving," "reflective," "sentimental," "tense," and "trustworthy"— and asked to choose the five or six you think best describe you. Then, your friends and/or colleagues are given the same grid and asked to select the five or six words *they* think best describe you. By overlapping the finished grids you can see what are often both striking similarities and profound differences in how you view yourself (e.g., "happy," "intelligent," "patient") and how others view you (e.g., "dignified," "reflective," "shy").

This personality-focused version of the Johari Window offers a model for understanding how we give and receive information. In terms of analyzing personality awareness, the respondent in the above example who thinks of himself as "patient" because he routinely counsels the junior members of his staff when they have made a mistake, may be shocked to learn that those junior members think of him as too "shy" to come right out and tell them how annoyed he really is when they make mistakes. The signals the person is sending out are being received in a completely different way than he imagined. It is to his benefit to become aware of the disparity between how he sees himself and how others see him; this increases both his and his staff's ability to handle business more effectively.

Now, as marketers, it is our ability to efficiently and precisely exchange information that shapes how efficiently and profitably we pursue our business. The management consulting firms that use the Johari Window employ it as the basis for improving this exchange of information—for restructuring the corporate communications processes. In other words, this test is adapted to map, not personality awareness, but *corporate* awareness. Assessing personality awareness can help a person to know more about how his actions are being interpreted by those around him. Assessing corporate awareness can help a business understand what it doesn't know about itself and its customers. Critically, this can assist a company in understanding how its actions are interpreted—and how these interpretations, positive and negative, impact its bottom line.

A simple matrix lays out all possibilities into four quadrants. The quadrants are labeled: "You Know You Know"; "You Know You Don't Know"; "You Don't Know You Know"; and "You Don't Know You Don't Know." All these "you knows" and "you don't knows" can seem both confusing and self-evident at the same time, so let us examine each quadrant individually.

You Know You Know	You Know You Don't Know
You Don't Know You Know	You Don't Know You Don't Know

YOU KNOW YOU KNOW

In the first quadrant, You Know You Know, are all the things that you, and others, recognize about yourself. In terms of personality analysis, this area can be a very public, often benign domain, including such information as the color of your eyes, the make of car you drive, whether or not you are married, and what your favorite basketball team is.

At first glance, it would seem that knowing your company's corporate address, its logo, and this year's marketing slogan would be the corollary to knowing such basic personal information as eye color. But in corporate terms, the content of this quadrant can be more ominous. What it really means is the things you and/or your company take for granted, things that you are certain about—even *smugly* certain about—as well as the things for which you rest on your laurels.

Take a moment right now to ask yourself the following questions: How long has my company been targeting the same market segment? When was the last time we scanned the horizon to see what, and who, was new out there? Have we been rolling blindly along, absolutely sure of who our customers are and forgetting *who our customers might be*? Is the idea toolbox we work with every day outfitted with the latest equipment, or are there still a bunch of old IBM Selectric typewriters in there?

It's a point we've already made (and because it is such an important one, you can expect us to make it several more times before you reach the end of this book), but one of the most broken-down old Selectrics many companies still sentimentally hold onto is the idea of a "mainstream market." Seth Godin, the author of many bestselling books about marketing, encapsulates the demise of the mainstream market with his admonishment to "think small." That is, marketers must stop regarding the people who use their goods and services as one throbbing mass of humanity, and start to consider each person making up the mass. Companies must start to think about how products and services can be made to resonate on a more individual level.

Three primary reasons for the demise of the mainstream—the reason that traditional print and broadcast advertising just doesn't work the way it used to any more—are, first, that there is so darn much of it. For decades, people have been bombarded with companies' "big ideas" to the point that they have learned, for the most part, to ignore them. Case in point: we were watching a basketball game with a few friends recently and the halftime commercials were hysterical. Everyone in the room was laughing out loud. This piqued our marketing instincts. We decided to conduct a quick survey. Pausing the DVR, we asked, "Quick—name one of the products being advertised just now." Not one of them could do it. It wasn't that the commercials weren't well done. With excellent production values and fall-down funny scripts, the commercials were the pinnacle of the form. But whether it was selling beer or cars or a new shaving system, no one could remember the brand—and sometimes not even the product. The commercials were a comic interlude to the main sporting event and, for most people who aren't in the marketing profession, that's all they were. Look at it this way: when a target customer sees your Web site, print ad, TV ad, brochure, direct mail piece or hears your radio spot, it may be the first time that person is encountering your brand or product. And the average person encounters 300 to 3,000 advertising impressions a day, depending on which study you believe and how you define an impression.[1]

That many impressions aren't just indicative of a saturated market—the consumers are also just plain soaked. No wonder people are growing more discerning about the impressions and images they allow to penetrate their consciousness. Now marketers are being challenged to find ways to impress consumers on a whole new personal level. Several years ago, a group of researchers at Harvard University asked, "How many times must prospects see a marketing message to take them from a state of total apathy to purchasing readiness?" After a year-long study, they returned with a definitive answer: nine times.[2]

The second primary reason for the change in how consumers absorb and react to media is, of course, the change in media itself. Internet, social media, and other advanced technologies allow consumers to "self-filter," if you will, the marketing impressions they encounter.

The myriad real, concrete ways in which one can attract a certain niche target segment is a subject we'll take up in much more detail—especially as it applies to AAAs and luxury products—later in the book. For now, let's simply acknowledge that luxury brands have customarily—and necessarily—devoted marketing dollars to appeal to a small niche of often hard-to-reach consumers: the affluent. At this juncture, the key is to recognize that the marketing and advertising landscape has changed. Traditional markets are shrinking, aging, or simply oversaturated. In order to broaden their customer base, marketers must look for new ways to reach new and ever more finely-defined audiences. They have to find more efficient uses for their money in this new landscape.

Larry Light, a former chief marketing officer (CMO) of McDonald's, is a visionary in understanding the growing importance of niche and target marketing. As early as 2004, Light put his finger directly on the radical shifts in media consumption that were changing the marketing landscape. In a speech to that year's AdWatch Outlook conference, he said: "Any single ad, commercial, or promotion is not a summary of our strategy. It's not representative of the brand message. We don't need one big execution of a big idea. We need one big idea that can be used in a multidimensional, multilayered, and multifaceted way."

The big idea, in other words, has to be able to be translated into several different media, and also needs appeal that can be translated for myriad different audiences. As Godin puts it, "Your ads (and your products!) shouldn't cater to the masses. Your ads (and products) should cater to the customers you'd choose if you could choose your customers."[3] Exactly!

Why not take a page from the playbooks of the best and the brightest and do a little exercise in choosing your customers? Imagine that you are the marketer for an upscale Maui resort. Among its many amenities, your resort features two professional golf courses that offer a challenging game and unsurpassed views that overlook the Pacific from some of Hawaii's most majestic shorelines. If you were to hear of a niche segment of golfers that had grown by 30 percent in the last decade, you'd naturally want to know who they were so you could tell them about your golf courses. You would definitely want to have this exploding new group as part of your customer base.

The African American golf community has, since 1996, increased by 30 percent. In fact, the National Golf Foundation reports that a full 15 percent of golfers—a stunning 5.5 million people—are minorities, and 2.3 million of them are African American.

Adding 2.3 million people to a list of potential consumers is a marketer's dream come true. And all one has to do, in the case of our beautiful Hawaiian golf resort, is let go of some past assumptions about the "typical" golfer and step into the modern golf market.

This brings us to a third key reason for the demise of the mainstream or the general market. However unconsciously, the term "general market" has long been a code term for "White," and these old assumptions are just not in sync with new demographic realities.

The bottom line is: don't always assume you know what you know.

You Know You Don't Know

The second quadrant, You Know You Don't Know, is perhaps our favorite one, because there is no better way to learn than by admitting you don't know something.

Granted, it takes a certain confidence in oneself to summon this kind of everyday curiosity. After so many years of being successful in the field of marketing it is difficult to look outside of one's experience and "fess up" when faced with a situation that tests the limits of one's knowledge. A friend casually uses a French phrase we're not familiar with and we laugh, hoping that this is the appropriate reaction, instead of simply asking for a translation. Our son asks for help with an especially difficult algebra problem on his homework and, rather than let him think that we don't walk on water, we soberly admonish him that he's got to figure out how to do it himself or he'll never learn algebra at all.

But in a business situation, knowing what you don't know and admitting it is the hallmark of an effective executive. It is impossible to thoroughly evaluate a competitive situation if you don't recognize the limits of your knowledge and make a plan to get the information needed to make a competent decision. We're pretty sure that when Los Angeles Lakers coach Phil Jackson evaluates his team's defense he doesn't do it in the dark. He gets tapes of the past games the Lakers played against the opposing team, tapes of the opposing team's games with every other team they've played that season, the input of his players and assistant coaches, and likely, has the latest injury report from the doctor who's been treating the torn ligament in Andrew Bynum's right knee. He is constantly evaluating and re-evaluating the playing field to get a competitive edge.

It's true that high-end brands have been searching out and acting upon innovative information for new ways to broaden their customer base and reach the various segments within it. For example, for several years luxury brands have been emerging with more affordable product lines to appeal to fresh market segments. Joseph Abboud is scheduled to launch a new line of his clothing at J.C. Penney, much as Isaac Mizrahi has already done successfully for Target. Outlet malls feature shops that sell such upscale brands as Gucci, Neiman Marcus Last Call, and Fendi. Merchandising in outlet malls is a smart move to reach both the mall's typical middle-market shopper while also providing an entry level to the brand for future, truly affluent customers. Also among the refreshingly different venues for shopping for luxury goods is, of course, the Internet. As Constance White, fashion expert and Style Director for eBay told us, "Shoppers who seek new designer fashions, limited edition, discontinued or vintage items in order to be distinctive find eBay to embody the ultimate luxury store. eBay allows our customers to virtually travel to Europe, New York City, or China without the cost."

There are whole new ways to shop these days, and luxury sellers are at the forefront of capitalizing on these opportunities in ways that still maintain the upscale identity of their brand. But whether the stores are brick-and-mortar or cyber emporiums like Red Envelope and eBay, how do marketers get the customer—particularly the AAA—into the shop? How do we provide a consumer experience that will turn them into comfortable, loyal customers who will tell ten of their closest friends about us? What information do we need to make that happen?

Ethnic minorities have been prime targets in the marketing efforts of mass brands for many years. Ford, State Farm, Pepsi, Proctor & Gamble, and McDonald's have all successfully made an impact in this segment. But let's be precise about the segment we're referring to. These marketing efforts have not been especially attuned to the AAA segment in question. The emphasis of these brands has been on the so-called "big numbers" represented by the large middle segment of this group—and, not secondarily, much of their marketing efforts have focused on employing a highly visible few Black athletes and hip-hop personalities as spokespeople. Of course, hip-hop has been incorporated for use in the advertising of luxury goods as well. But is this the best decision a brand can make? Gucci is a prime example with their new advertising campaign featuring pop star Rihanna.

In these ads, Gucci is doing what experienced marketers have traditionally done: reach out to new prospects by embracing a new segment, whether that segment involves minority, crossover, or youth-oriented consumers. The Rihanna ad attempts to reach all three. However, the reality is that the vast majority of the significant affluent African Americans niche is not part of hip-hop culture. Yes, hip-hop culture transcends race, religion, creed, and age, but it's not a magic bullet for reaching the AAA audience.

In an article for the *Atlantic Monthly*, journalist Hua Hsu wrote of hip-hop's influence:

> Over the past thirty years, few changes in American culture have been as significant as the rise of hip-hop. The genre has radically reshaped the way we listen to and consume music...[but] hip-hop is more than a musical genre: it's a philosophy, a political statement, a way of approaching and remaking culture. It's a lingua franca not just among kids in America, but also among young people world-wide. And its economic impact extends beyond the music industry, to fashion, advertising, and film.[4]

This is all true enough. Indeed, ten or fifteen years ago, many AAAs helped shape hip-hop culture. In the 1990s, luxury brands bought into the power of "urban"—cars, clothing, jewelry—by embracing hip-hop and rap culture. But fast forward ten years. What do you think has happened to these consumers? They are no longer 18 to 34 years old, but 28 to 44 years old. What are luxury brands doing to appeal to this audience now that they are all grown up? A few are taking advantage of the marketing opportunities built into this segment—think Louis Vuitton's design partnership with Pharrell Williams or Gucci's philanthropic partnership with Mary J. Blige.[5,6]

But too few understand that this now mature audience is still aspirational. What does that mean for the luxury retailer?

> These days customers are finding it far harder to forget about price. The seriously rich, of course, are still spending freely. But much of the industry's rapid growth in the past decade came from middle-class people, often buying on credit or on the back of rising house prices. According to Luca Solca of Bernstein Research, 60% of the luxury market is now based on demand from "aspirational" custom-ers rather than from the wealthy elite.[7]

But this consumer is not merely aspirational. He now has more access, money, and sophisticated tastes than he did back in the 1990s.

The key concept for marketers to remember when thinking about this maturing urban segment is "ten or fifteen years ago." As the bobbysoxers of the 1940s grew up to be the suburban matrons of the 1950s and 1960s, or as the revelers at Woodstock in 1968 grew into the businesspeople and entrepreneurs of 1988, the members of the urban youth culture of 1999 are older and wiser now. Their tastes, and their pocketbooks, have matured. Why not finish what was started when this group was just starting out—and expand on it?

Marketers need to look beyond the hip-hop reference point when planning media strategies. While Rihanna may make some potential AAA customers give the glossy ads a second look, according to our research, her image is not a purchase motivator for this prosperous audience. Their lives are as far removed from Rihanna's as any adult life is from that of a teenager's. Just as any well-heeled adult is likely to be unresponsive to a call to the good life from, say, Britney Spears or Lindsay Lohan, AAAs have no aspiration to be like—or, critically, even *seem* to be like—an eighteen-year-old music star. It's not that they don't like Rihanna or think that she's talented; it's that she's just not on their radar.

In order to effectively market to, and ultimately win, the affluent African American customer, it is important that companies demonstrate in their communications, marketing programs and sponsorships, an understanding of and respect for both African American culture *and* the affluent segment in particular. A well-known credit card company managed this very well with its memorable "Are you a card member?" campaign. With a constellation of African American movie, music, and sports stars from which to choose a spokesperson, the credit card company featured the prominent filmmaker Spike Lee reporting that he had been a member since 1985. The choice of an elite member of the African American segment is crucial to the AAA's ability to identify with the brand. Gucci might have made a better choice in terms of reaching the AAA audience by choosing to employ an image of Halle Berry, Zoe Saldana or Thandie Newton, or even an undiscovered Black model from Africa in their ads—in other words, choosing someone who registers on the AAA radar, someone who inspires their aspirations. For example, look at what a virtually unknown Black model by the name of Tyson Beckford did for Ralph Lauren in terms of appealing to the AAA segment.

The subtle misstep of the Gucci ads is not an isolated example, either. These mistakes are nearly everywhere we look. On a recent stroll down New York City's Madison Avenue, we passed a luxury boutique. In the window was a mannequin, dressed in the store's swank apparel. Nothing remarkable there. But this mannequin was Black and had dreadlocks. In the typical AAA environments—corporate offices, Wall Street, country clubs—there are very few dreadlocks flying about. The mannequin was yet another example of trying to make a brand seem "cool" by associating it with urban culture, or the result of a company making an honest effort to reach out to a particular luxury consumer but missing entirely the nuance that would make this consumer truly connect to it. Maybe the shop was vying for the attention of a younger demographic. Regardless, the bottom line is that a Black mannequin alone—even one sans dreadlocks—will not increase sales. There needs to be far more in the marketing mix than that.

According to Greg McBoat, chief economist for the marketing consultancy Diversity Affluence, research consistently confirms that AAAs are ready, willing, and able to purchase what luxury brands have to sell. But few marketers know enough about this segment to fashion campaigns that appeal to them directly, nor how to execute those campaigns using the appropriate media and marketing outlets.

The second quadrant, then, is filled with all of those bits and pieces and piles and mountains of information we know exist but do not have at our fingertips. Operating without the benefit of appropriate information can cause the sort of missteps we've just talked about—and worse, in the emerging twenty-first century marketplace—could cause even greater damage to a brand than the embarrassment and expense of a simple misstep. When a marketer approaches a new segment without really knowing much about it, he or she runs the real risk of alienating that market. This disaffection can result from either passive mistakes ("We've got that segment covered in the general campaign; we don't need to do anything special.") or active ones ("Let's just put some Black entertainer or athlete in the creative; that'll attract the Black audience."). This sort of self-defeating thinking might also be the result of the sort of laurel-resting we discussed earlier ("It's not that big of a consumer group so who really cares?" Or, "We've got this segment and now we don't need to place any more money against this audience.").

Possibly even worse is the risk of actually attracting this audience and then undervaluing them. By undervaluing them we are referring to attempts at diversity marketing that are not real strategies or thoughtful programs, but one-shot deals—a few print ads, one big event—that can make a splash but lack the consistency to keep a brand at the forefront of the consumer's conscious.

Here's a fact for you: the total Black population in America spends approximately $279 billion annually on consumer goods, but advertisers spend only about $865 million to reach out to them.[8] We have to believe that's because some of these advertisers don't know what they're missing—or that some of them simply have no idea how to execute campaigns that would resonate with the Black audience. And others, frankly, will never do this because they don't want their brand becoming "Black." We actually once heard an executive for an auto brand tell us that they didn't want to market to women because they didn't want their brand to become a "chick" car. In this day and age, does it really make sense to ignore whole segments in this manner?

The saving grace, however, is that we *can* get our hands on the facts and figures we need to shape suitable, simple, affordable, and nuanced marketing strategies to attract a particular consumer segment. Indeed, the facts and figures you'll need to successfully tap into the AAA market are in your hands right now. Clearly, since you've picked up this book, you know you don't know everything you need to know to reach this audience—and knowing you don't know is a very smart first step.

YOU DON'T KNOW YOU KNOW

The third quadrant, You Don't Know You Know, is where some mystery lies among the countless bits of information at your fingertips that are implicit, but rarely acknowledged outright.

Suppose you own a restaurant. It's a white-tablecloth place with an extensive list of fine wines and a chef whose celebrity is growing. As the owner, you are keenly aware of the costs you incur every night. You know that you burn through twenty-six dollars a night on the white candles lighting the tables. You know your linen service charges sixty-four cents to launder every four-top tablecloth and nine cents for every napkin. You know that it costs you eight dollars to plate every filet mignon you serve to a customer, and three dollars to plate every lamb shank your chef ran as the special last weekend. And you also know that somehow you simply have to cut costs because, while most restaurants turn a profit of between 5 and 7 percent after expenses, you are turning a profit of only 3.5 to 4 percent. So you talk to your linen service rep to renegotiate the cost of laundering. You comparison shop among restaurant suppliers to find less expensive candles. You may even consider raising your menu and wine list prices by a percent or two in the effort to bridge the gap and run a profitable enterprise. The most obvious route to a bigger profit margin, however, may not be the first to strike you. Take another look at your plating costs. The difference between plating a filet mignon and a lamb shank is significant but—and here's the key—the price on your menu for the filet is thirty dollars and for the lamb shank is twenty-eight! If you serve a table-for-four steaks, your gross profit on the food is eighty-eight dollars. But if just one person at that same table-for-four orders the lamb instead, you've immediately increased your profit on that table by a whole percentage point. You're an awfully smart restaurateur; your budding celebrity chef's special lamb shanks are going to become a regular menu item.

So, what is it that you know, but may not be allowing yourself to acknowledge about the AAA market segment? Take a step back with us for a second and we'll tell you.

In spite of the impressive and growing numbers of AAAs, they are not yet a sought-after market and partly this is because affluence is still seen as the exception rather than the rule for this segment. How often does a marketer react to the sight of a Black man in a Ferragamo or Nedo Bellucci custom-made suit behind the wheel of a BMW and think, "Wow, there must be a whole lot more affluent African Americans out there!" Probably a lot less often than he or she thinks, the person in the driver's seat of the BMW somehow does not fit into the general, accustomed pattern.

One of the reasons for this may be that, while marketers and brand managers might be knowledgeable about their product category, they are just as influenced as the rest of society by an overflow of media images that portray Blacks as part of an uplifting history lesson, or survivors of a modern-day struggle. But the stories of African Americans are as richly diverse as those of any other ethnic group. Most AAAs did not rise from a racial struggle in the one generation removed from the civil rights era but from a Black middle class that has been thriving since just after the Civil War.

Are you surprised to know that there has been a thriving Black middle class for almost 150 years? What's important for marketers to understand is that while AAAs are very proud of their heritage, they are as removed from the struggle as the Kennedys were removed from the days of "No Irish Need Apply"—and so are very unlikely to be defined by it. The AAA life story isn't limited to the themes usually found in the mainstream media. And neither are their aspirations.

Think again of the Gucci campaign featuring Rihanna. One could look at this ad, featuring a beautiful Black woman dressed in designer fashions, and think of it as a positive positioning of Blacks in America. Without a doubt, this is a step forward from the way Blacks have long been portrayed in the media, advertising in particular. Think of Aunt Jemima, Uncle Ben, and Rastus, the Cream of Wheat chef. Indeed, however protracted and ongoing the effort to forge new and more accurate images of contemporary Black life in the media, approaching any target audience with respect has to be a no-brainer for marketers. In the post-Obama age, aren't we beyond the point where we can pat ourselves on the back for portraying any ethnic group with simple dignity? Positive images on their own are no longer, as if they ever were, enough.

Now let's look at another key word in that last paragraph: "contemporary." What are the contemporary concerns of AAAs as a community? Well, in a way, that would be like asking about the contemporary concerns of Irish Americans and expecting one person who happens to have the surname Murphy to give a blanket answer that applies to every member of the entire group. Irish Americans aren't monolithic and neither are affluent African Americans. Similarly, the issues, and even the struggles, that characterized their grandparents' and parents' generations are no longer primary concerns for these successful professional people. For AAAs, the United Negro College Fund, the National Association for the Advancement of Colored People, and other civil rights era institutions, while still very much honored and supported, do not represent the pinnacle of either social networking or charitable giving. A very proud heritage is gladly acknowledged, but this is not the sole platform on which the AAAs live their lives.

Paradoxically, this is exactly the customer profile that can explain why some luxury marketers approach this target segment in such a laissez-faire manner. "Oh, this is just another kind of luxury client. We'll be able to reach them with our general marketing because, once people get to a certain level of wealth, they assimilate."

This assumption is true, but only to a degree. Even though AAAs may have more in common with what you view as your core customer base than you originally thought, they also constitute a separate opportunity for several unique reasons: their growing purchasing power, purchasing habits, and cultural traditions. The general market has had little impact in this segment just because these unique reasons are not widely understood by marketers. By reaching the traditional luxury market you may be able to sell the AAA something once, but you won't become part of their lifestyle. You won't turn them into a brand loyalist. For that, you need a deeper understanding of this market on both a cultural and historical level, and you'll need to find out what sets them apart from your existing market. Because while we acknowledge that AAAs have a lot in common with the general luxury market, it is the ability to understand how to reach them that will help us create a marketing plan and a loyal customer following among them. Think, one more time of the Gucci ads that featured Rihanna. In what print venues did this ad appear? *Vanity Fair*, *Vogue*, *Glamour*? Yes, the ad was in all of these publications. The publications that the Rihanna ad missed? Magazines like *VIBE*, *Uptown*, *Essence*, *NV*—magazines, or event fundraising event programs, that are targeted specifically to the African American audience.

There are several reasons that integrated marketing campaigns aren't executed and advertisements for luxury goods don't show up in Black media. The first is that luxury marketers often don't know what is available to them or don't understand the impact of advertising in Black-oriented publications or forging marketing partnerships, or they make assumptions about the readers of that publication and so miss the opportunity to reach these readers, thinking they'll just catch them when they pick up that month's *Vogue*. But a little-known fact about Blacks and Black-oriented print media is that magazines are 6 percent more important to this audience than they are to the population at large.[9] African Americans, and in particular AAAs, are loyal readers of publications that embrace their affinities, activities, and aspirations. What is likely even a broader revelation—though it will seem wildly obvious once we've said it outright—is that as this audience also consumes general market media, *it notices when the luxury brands that fill the pages of* Vogue *and* Vanity Fair *are absent from ethnic media*. AAAs are very discerning shoppers and can easily perceive whether or not a firm, corporation, or store is welcoming. Research has shown that they select who provides their goods and services based on the amount of respect shown by these entities for their needs. They prefer to consume media that is culturally relevant and features creative designs that include and/or reflect them and images important to them. It behooves marketers to know what these ethnic publications are, and in the chapters that follow we're going to outline exactly what's available and talk about the significant amount of money that is being left on the table in bypassing these venues.

By far, however, the most broadly used excuse for a lack of presence in Black publications is, "We don't have the right creative," so let's talk a little about that right now.

Often, when people of color are shown in the creative, it is to evoke the feeling that the product is one that is cool and stylish. It's almost an equation in the fashion world—people of color = trend-setters—and, if you want to be a trendsetter too, you'll buy this jacket, watch, car, or whatever. But when the creative is specifically slated for publication in a Black-oriented venue, we hear marketers dialing back: "The creative's not cool enough." They believe the creative won't "fit" in a magazine such as *Essence* and are sometimes reluctant to spend the money to create an ad specifically for this venue.

In fact, research shows that you don't always need a new creative to market in these venues. You can use in *Essence* or *Uptown* the very same ad you used in *Vogue* or *Vanity Fair*.

The ad doesn't have to be any "cooler" than the one you're proud to run in any other major publication. And, here's a shocker, the ad doesn't even need to have any Black people actually in it. It's nice to see images that reflect you and your lifestyle in advertisements, but a good ad is a good ad, no matter what.

It's much more important that the advertiser show up in media that you participate in.

Let's take a hypothetical situation, because we don't want the simplicity of that last statement to get lost. Suppose you have a company that makes skateboards. You are probably going to advertise in magazines like *Skateboarding, Concrete Wave,* and *Freestyle.* You are probably going to sponsor skateboarding events like the Dew Action Sports Tour and local skate days, and you are probably going to attend industry events like the Action Sports Retailers convention. You don't waste your time and money taking a booth at the state flower show, and you don't buy a lot of space in the *American Association of Retired Persons (AARP)* monthly publication because there probably aren't many orchid aficionados and/or retired people who ride skateboards. And you aren't going to rely on ads in general youth magazines in the hope that these magazines have some skateboarding readers. No, if your company makes skateboards, you go to the places where the density of skateboarders really is. You are going to go their playgrounds because you want to make your product relevant to their lifestyle affinities, activities, and aspirations.

If you want to reach affluent African Americans then, it is a matter of making your product relevant to their lifestyle, affinities, activities, and aspirations. Show up in the magazines they read, and at the places they congregate, and at the events that have meaning and value to them. This is "lifestyle marketing," something with which luxury marketers are very familiar because it is a critical strategy. Some companies, like Mercedes, have made an art out of lifestyle marketing. Finding out where customers play and going to their playground has long been something marketers know they must do. Now you just have to apply that knowledge in your quest to attract the AAA market, too.

YOU DON'T KNOW YOU DON'T KNOW

The most deadly quadrant is the last one—You Don't Know You Don't Know.

In a social situation, people who carry on when they don't know what they don't know can be boring and annoying. In a business situation, not knowing what you don't know is just plain dangerous. At one point we were helping an upscale island resort fine-tune its marketing efforts. The resort had everything a luxury vacationer desires: beautiful hotel rooms, first class restaurants, luxurious shops, several pools, and white sand beaches to boot. Business at the resort was steady, but the managers felt that it had hit a plateau. They wanted to shake up their marketing efforts, figure out what segments they might be missing, and reach those segments.

Upon studying the photographs of the resort, our first comment to them was, "There are no Black people here."

Well, no, there weren't, and the resort's executives weren't surprised about that because they had never tried to market the place to African Americans. Never even considered it.

There are a host of reasons why the AAA segment isn't recognized as a vital one in the offices and conference rooms where marketing decisions are made. First, in some ways, it is an invisible market. As we have mentioned, when marketers have traditionally thought of affluent African Americans, their first point of reference is often Black celebrities and athletes. This isn't necessarily an uncommon failing—65 percent of the entries on *Forbes'* recent list of the top twenty wealthiest Blacks in the United States *were* athletes and entertainers, Oprah Winfrey and Tiger Woods and Michael Jordan and Jay-Z, to name a few. But the other 35 percent were people like Janice Bryant Howroyd, Quintin Primo III, Herman J. Russell, Alphonse Fletcher, Don Peebles, John Thompson, and Sheila Johnson—people who'd made their fortunes in fields like finance, employment services, construction—and the names of these Black billionaires are, unfortunately, not on the tips of the tongues of the general public. When marketers think of African Americans, they think of entertainment and athletics as the only models of Black success. Their blind spot is the Black doctors, lawyers, entrepreneurs, and others in corporate America whose professions are what support the $87.3 billion in buying power we're writing about in this book. Brooke Astor and Princess Grace were not the only wealthy Whites of a previous era, and Oprah and Jay-Z are not the only wealthy Blacks of ours—it's just that, like Astor and Grace, they are the most notable.

A second reason the AAA segment is overlooked overlaps a bit with the first. Marketers actually recognize this segment as a viable one but, because they don't know the *extent* of its viability (i.e., its growing numbers or the amount of dollars it controls), it is far down on the totem pole of urgent things for them to consider. These people are leaving at least $87.3 billion dollars on the table because they haven't done their homework. Because they don't know what they don't know.

At the other end of reasons why the AAA market isn't targeted is—and this is pretty amazing—that the company does not want this audience. They fear it will have a negative impact on the brand's stature or heritage and Whites may not want to participate in the brand if it is taken up by Black consumers. Of course, that company's executives are not going to come right out and tell you that's their reason. We worked with an apparel company at one point and their rationale for not targeting the AAA market was that their clothing was not "urban." It was not "hip-hop." It was made for the "outdoors."

"Do you want Black people on the island?" we asked the resort's executives, jumping right into the kind of candid, and frequently awkward discussion that can be a part of a healthy dialogue to help a company redefine and broaden its marketing goals. "I mean," we continued, "are they not coming because you're not marketing to them, or do you assume they're not coming because they don't want to be here?"

One of the marketers in the meeting filled in the uncomfortable silence that followed our questions with the comment, *Black people don't like to swim.*

"Do you have research to support this fact?" we replied.

Now things grew even more awkward. There was no research. The marketer had based his entire AAA strategy on a remark he'd "kind'a heard somewhere once."

The positive outcome of this particular awkward conversation was that all of the executives were candidly innocent, so the exchange was a refreshing one with people who were open to learning what it was that they did not know. But can you imagine the sort of unwieldy marketing campaigns—and ravaged marketing budgets—that would result if all of our marketing decisions were based on conjecture?

Now, by the way, making decisions on statistical analysis alone is also conjecture. Statistics need to be interpreted in order to understand the nuances of this audience and craft an effective, authentic message. But it's just too obvious that it's bad business to base a corporate strategy on remarks overheard in a locker room or, possibly worse, things that got said in passing over a family dinner table when we were kids that we absorbed as fact and never questioned. Retaining these perceptions from years gone by is what is known as "taxing"—whereby you "inherit" your distorted perceptions—and that's the place from which stereotypes emerge. And in corporate America, thinking like this is only an opinion unless it is supported with cold hard research. We need to turn a critical eye on the things we "kind'a" heard somewhere—*Black people don't like to swim*—on blanket statements that are patently absurd when held up to even the slightest ray of light.

"Do you want Black business that's currently projected at $87.3 billion in purchasing power?" we asked the resort's marketing team. Of course they did. And all they really had to do to pave the way for developing an effective strategy to get that business was to get their own misperceptions out of their own way.

In the twenty-first century, Obama-is-president world, the entire market is undergoing a paradigm shift—a change from one way of thinking about business, and people, to quite another. The shift has already begun in the way marketers are viewing, and approaching, emerging market segments. In 2004, Chris Anderson coined the phrase "the long tail" to refer to niche strategies that sell large numbers of unique items, each in relatively small numbers. But successful niche and target marketing requires seeing people who you think you know in a different, and more accurate, light. When this happens—when a marketer can see the AAA niche for what its potential really is—it parallels that "Aha!" moment after the Iowa caucuses when people realized that Barack Obama was a candidate who had to be taken seriously. And what that Aha! will lead to is the same sort of understanding that enveloped the world after Jackie Robinson entered the major leagues. Marketing now, like baseball then, is a completely different game.

Unfortunately, most luxury brands have yet to reach their Aha! moment. Or, if they have, they do not yet know enough about the AAA segment to know what to do with it. In a new study of seventy-four marketers from member companies of the Association of National Advertisers, over three-quarters (76 percent) have programs that focus on the African American audience.[10] But just 45 percent of those respondents indicated satisfaction with the results of their multicultural marketing initiatives, and a full 26 percent said they were either only "somewhat satisfied" or "very dissatisfied" with the results. They cited several key obstacles to satisfaction, including that only a quarter of them feel that their firms have a high degree of knowledge and disciplined best practices when attempting to appeal to ethnic consumers. Many indicated that they were frustrated in trying to integrate diversity programs into their overall marketing mix. And, while more than half of those queried (57 percent) were aware that successful niche marketing requires separate messaging for distinct market segments (and communicating this message via media that directly reaches the multicultural consumer), more than 20 percent are still relying on outdated and, frankly, less sophisticated marketing strategies. Eleven percent report using a "mainstreaming" strategy—a repurposing of general advertising approaches to appeal to the multicultural market—and 10 percent simply translate general marketing materials for media catering to multicultural audiences.

This repurposing and simple translating of general advertising is rather akin to the *New York Times* position of being "platform agnostic"—simply transferring the information that traditionally came to us in ink on newsprint to digital. The *Times* is losing readership because it hasn't broken new ground in the digital format that people are increasingly relying on for news—namely, the Internet. Similarly, marketers who aren't breaking new ground in terms of how they approach niche consumer audiences aren't any more likely to attract new customers within these niches than the *Times* is attracting new, Net-savvy readers.

Now, let's talk briefly right here about what we mean by "breaking new ground" and making inroads with the AAA segment. We do not mean that you need major media to reach this audience—you don't always need to hire an agency or spring for a flight of television commercials. If you do have an agency or plan to hire one, it's likely they aren't leveraging their human and intellectual capital to the brand's advantage. The strategies we are going to propose to you in this book can be grassroots and affordable. They are so simple and available they might strike you as *too* simple. Well, maybe marketers have made marketing too complex and it's time to get back to the basics.

What you want is for the members of the new generation of affluent African Americans to identify with your brand or, better yet, to have your brand identify with them. By penetrating African American households with minimum incomes of $150,000 or more, you will earn an open invitation from a whole new loyal, upscale audience and their circle of influence. In order to do this, you need to stretch your mind past old and outdated reference points for this segment. Put aside what you think you know and follow us into the next chapter: a brief history of where the AAA market really came from, and how it has emerged to the current moment of its growing significance. This short narrative may surprise you, but you will never again put Black history in a nutshell. This brief background will also give you the insight you'll want to really understand how we are going to change the old game, and how bringing back simple and affordable marketing basics can really resonate with your bottom line.

CHAPTER 2

THREE DEGREES OF SEPARATION: EVERYONE KNOWS EVERYONE

No marketer expects that any campaign he devises will have a perfectly linear impact in the marketplace. That is, that one print ad, radio spot, or special event will influence one potential consumer in isolation from every other potential consumer—or in isolation from the culture at large. Indeed, no marketer would *want* to take such an inefficient approach. For decades, it's been a mark of success when an advertising campaign becomes a part of the culture. Think, "Where's the beef?" or "I can't believe I ate the whole thing." These are certainly broad, mass market examples. And they may seem dated, but we use them here purposefully to make the point that ad campaigns could go viral even in the days before we had the technologies that put the phrase "going viral" into our vocabularies. That's because "going viral" has always had a unique, person-to-person, word-of-mouth component.

THE HUMAN WEB

Most of us are familiar with the popular trivia game "Six Degrees of Kevin Bacon." The idea of the game is to link the actor Kevin Bacon through his film roles with any other actor, and to do it in six moves. For example, Will Smith was in the film *Welcome to*

Hollywood with Cathy Moriarty, who was in *Digging to China* with Kevin Bacon. Therefore, Will Smith has what is known as a "Bacon Number" of two. On the other hand, Meryl Streep starred in *The River Wild* with Bacon, so she has a Bacon Number of one.

But the real-life Kevin Bacon has also turned what one of its inventors called "one of our stupid party tricks" into a way to do something good. As the actor says of his Web site, SixDegrees.org, where visitors can support their favorite charities or those of their favorite celebrities, "You've probably heard of the Six Degrees concept. Any one person (including me, Kevin Bacon) is connected to any other person through six or fewer relationships, because it is a small world. SixDegrees.org is about using this idea to accomplish something good. It's social networking with a social conscience."

Bacon is well aware that he is playing off an idea that was around long before he was born. It is the theory of "six degrees of separation," or the "human web." This theory, popularized in a 1990 play by John Guare, says that every person is one degree away from each person they know, and two degrees away from each person who is known by all of the people they know, and that ultimately, every person in the world is only six degrees away from knowing every other person in the world—the trick is in finding the right six people.

In the AAA community, however, the human web is even more densely woven than that. Among college-educated, affluent African Americans that degree is actually three. It's not a small community—let us remind you that we're talking about 343,300 households in just the United States enjoying a minimum annual income of $150,000 and 819,700 individuals who earn a minimum of $75,000 or more a year. But it is a tightly knit one. Now, it's likely impossible to figure out what everyone's relative Bacon Number would be in an AAA version of the game—even in the Kevin Bacon universe some actors have a number of "infinity" because there is no known way to link them back to Bacon—but our best and highly informed guess is that most AAAs are separated by only three degrees.

What do those three degrees mean in practical terms? How tightly and well connected is this group in real-world terms? Well, suppose we were sitting over dinner with a group of friends in Atlanta—or Los Angeles, New York, or Houston, for that matter—and one of these friends was volunteering, say, in a program to improve the quality of school lunches in her local public school district. This friend mentions that the children have planted an organic garden on school property and that part of their curriculum is designed around tending the garden, cooking with the vegetables they harvest, and learning about nutrition. She's very proud of this innovative program, but—here she sighs wistfully—it would be the pinnacle of accomplishment for these kids, not to mention public relations gold, if the school somehow got on Michele Obama's list of public appearances.

This isn't such a stretch when you consider that kids, education, and healthy nutrition are all initiatives the First Lady has taken to heart and made a part of her portfolio. But how does one go about directing Michele Obama's attention to one school's pet project when every other school in America would also be thrilled with a visit from her? Well, over dinner with a group of AAAs, it would not be unusual in the least for someone at the table to have gone to school with someone on Mrs. Obama's staff or to have worked with an attorney who worked with an attorney in Mrs. Obama's old Chicago office, or who had volunteered with the Obama presidential campaign and still maintained a friendly relationship with Mrs. Obama herself. Call it "Three Degrees of Michele and Barack Obama," but the odds are good that if you're at dinner with a group of AAAs, someone in that group is going to be able to pick up his or her cell, make a few phone calls, and end up on the line with someone who works in the First Lady's office. *That's* how tightly connected this group is. It's impossible to be better connected than that.

What we want to do in the remainder of this chapter is to show you how these long-standing connections developed and have been nurtured in a 150-year evolution to affluence that the AAA community now enjoys. We use the word "evolution" intentionally, for while most of us are familiar with the "revolution" begun by the civil rights movement, the ascent we're going to talk about has in great part taken place more silently and thoughtfully than the word "revolution" implies.

THE KEY TO SUCCESSFUL NICHE MARKETING

If you've gotten into the habit, as many of us have, of frequently conducting business lunches in the same restaurant, it's likely that the maitre d' knows you. Even if the restaurant is crowded, he will go out of his way to find a table for you, and, he often knows what you drink and asks if you'd like "the usual." Because you are a valued customer, the maitre d' makes it his business to know your preferences.

Naturally, the maitre d' takes care to remember what his regular customers drink because he wants to sell them that beverage and turn a profit on it. Just as naturally, the customer appreciates the effort he extends to make sure his or her dining experience is not only pleasurable, but personalized; indeed, it is probably, in part, exactly because the staff at the bistro know your tastes and take care of you that you have become a regular customer. The maitre d's effort, which, at the end of the day, is neither costly nor time-consuming, makes for a win-win situation.

Knowing who your customers are and what they like, and taking care of them is the key to successful niche marketing. As marketers, we know instinctively that success with any target demographic depends upon how well you know that segment and are able to connect with them in ways that resonate with their lifestyle.

What we may not always be aware of is how *increasingly* vital it is for the health of our brands to refine our concepts about what constitutes a new niche—and to know that niche as well as the maitre d' knows his guests' palates. So, how well do you know the AAA segment? How far removed are you from stereotypes that limit your understanding of this potential customer base? Do you have the kind of grasp of the AAA experience and perspective that will allow you to make culturally competent business decisions in marketing to this group? We spoke to Marc Bland, the manager of Analytical Solutions at R. L. Polk & Co., who puts it this way:

> As with most buyers, product and price are key, but to attract and retain the business of the affluent African American community...it all comes down to respect. It's important to respect the struggles and obstacles made to achieve their current degree of success. And it's important to acknowledge the gained earning power as well as their individuality and style preferences. These factors really need to be integrated into the advertising, sales, delivery and customer service aspects of a firm's business to genuinely appeal to this affluent community.

According to Robert R. Shullman, president of Ipsos Mendelsohn:

> There are a good number of market segments in the affluent and luxury markets that Mendelsohn has been measuring for the past thirty-three years in its syndicated media survey, The Mendelsohn Affluent Survey. As just one indicator of the importance of the higher income, African American market, in our 2009 questionnaire Mendelsohn is now measuring for the first time the audiences of upscale African American publications based on the requests of some major media agencies. There are over one million affluent African American households in the United States according to Mendelsohn's definition of affluence (household incomes of $100,000 or more) and these households spend billions and billions of dollars. As such, they are a very desirable target market for upscale products and services. Knowing who they are, what they buy and use, and how to reach them effectively is a critical issue for all marketers of upscale and luxury brands.

We think it's important to offer you an overview of AAAs, who they are, where they came from, and how they have risen to their current prominence. Again, we'll paint our portrait with broad strokes—this is by no means a history book. But, by and large, the current understanding of the AAA community is rooted in a collective history lesson we all *think* we've learned. As the American novelist and cultural critic, Touré, put it in a review for the *New York Times Sunday Book Review,* "For so long, the definition of blackness was dominated by the '60s street-fighting militancy of the Jesses and the irreverent one-foot-out-the-ghetto angry brilliance of the [Richard] Pryors and the nihilistic, unrepentantly, new-age thuggishness of the 50 Cents," that a look back is necessary to help us to "reset" our understanding of where this emerging class of affluent African Americans really came from.[1]

THE 150-YEAR ANNIVERSARY OF THE BLACK MIDDLE CLASS

The first insight is one we've already touched on: that the now-thriving Black middle class—not to mention the Black upper-middle class—did not spring solely and fully formed from the events, or the legislative progress, of the 1960s civil rights era. The fact is that there has been a vital Black middle class since the mid-1860s.

Granted, in the mid-1860s, this middle class was very small. While now nearly half of the Black population is middle class, in the years immediately following the Civil War, most Blacks lived in what could quite accurately be described as extreme poverty.

It is also about this time that subtle class distinctions were becoming evident within the Black population. For example, although most Blacks were poor, they still needed the services of teachers and nurses, hairdressers and funeral directors; they still needed places to shop for cloth to make their clothes and food staples to stock their pantries. They still needed places in which to worship. Because of either strict segregation laws or ingrained regional custom, they could not easily avail themselves of such services and products in the shops or within the organizations that served the White population. In order to fill this need, a class of Black merchants and shopkeepers, ministers, and other professionals in fields such as education, medicine, and law developed. In relation to the standards of living at the time, the Black middle class naturally still lacked the resources to afford many of the niceties enjoyed by their White counterparts—a White minister, for example, was almost sure to have a bigger church built from stone or other more substantial material than a Black preacher, or a White teacher was likely to have a greater number of more up-to-date school books than a Black one—but the status and comparative rewards conferred upon each within his or her own separate community was much the same. Coming in an era that followed close on the heels of one in which it was against the law for a Black person to learn to read and write, the Black people who possessed these skills were looked up to. Black teachers were especially honored—and the value of education became rooted within the consciousness of the community.

THE BIRTH OF BLACK URBAN CENTERS

The great majority of the nascent, free Black communities were born in the rural south. It wasn't until the Industrial Revolution began in the 1880s that significant Black urban centers began to form, and it was still later, during the "Great Migration" of 1914–1918 when many poor, rural Southern Americans, both Black and White, headed to urban centers in search of employment opportunities and better lives, that these centers began to grow and thrive.

Of the vibrant Black neighborhoods that began to grow in the northern cities, many of us are familiar with only the most famous ones—the Black Belt of Chicago, the chain of neighborhoods on the city's South Side where Michele Obama grew up; the now-demolished Black Bottom in Detroit; and New York's glittering Harlem, the center of the New Negro Movement, or what is now known as the

Harlem Renaissance. Like most minority groups at the time, Blacks lived in clusters, and the communities that emerged around these clusters were for the most part a response to the simple fact that, through law or custom, Blacks could not live and socialize in White areas and institutions.

In fact, however, the segregation was even more institutionalized. The time between the end of the Civil War and 1915 was known as the Age of Booker T. Washington. Washington was an American educator, orator, author, and the leader of the national Black community. He founded the Tuskegee Institute, a college that trained Black teachers, and made it his base of operations. In a famous speech at the Atlanta Exposition in 1895, he appealed directly to middle-class Whites across the South, asking them to give Blacks the chance to work and develop separately and, in exchange for this chance, he implicitly promised White leaders not to demand the vote. Most of the White middle class, as well as the Black middle class, were eager to follow his lead. Within the framework of Washington's bargain, Black communities began to flourish across the country—and within them, a tradition of self-reliance and self-help.

SALT LAKE CITY, UTAH, AND THE BLACK PRESS

It might surprise you to know that some of the nation's most active Black communities at the time were based in Utah. In fact, the Black population in Utah grew dynamically between 1890 and 1910, nearly doubling in size as those decades unfolded. Salt Lake City was the focal point of social, business, and political activities for Black Utahans.

The first institution usually established in these emerging Black communities was almost invariably a church. These churches—such as Salt Lake City's venerable Trinity African Methodist Episcopal Church, now a landmark on Martin Luther King Boulevard, and Calvary Baptist Church, which first began holding its meetings in

1896 in the back of a White church that had been set aside for Black worship—were the hub from which the organizations giving meaning to the lives of those in the Black community flowed: fraternal organizations, literary societies, Masonic lodges, and social events like parades, balls, and musical programs. Importantly, Black newspapers were also a part of the activity that hummed around these church centers. Some of the earliest Black newspapers were published in Utah.

The Black press, naturally, focused on reporting the ideas and ideals of their readership, and one of the things this readership was most interested in was injustice at both the local and the national level. In reporting news of national significance to the isolated Black communities, the newspapers began to weave the connections that helped the larger African American community develop a national network. It was, in fact, the two Black newspapers in Utah that had the most profound, and too often overlooked, impact on the political traditions that would emerge in the national Black community in the waning years of the nineteenth century.

The *Utah Plain Dealer* was published by William Taylor, and its chief competitor, the *Broad Ax*, was published by Julius Taylor (no relation). Both men were active members of the Western Negro Press Association (WNPA). It was a meeting of the WNPA in 1898, attended by both publishers, that set in motion a historic shift in Black political affiliation. Most Blacks at the time were loyal members of the party of Abraham Lincoln—in Salt Lake City, most of the political activities of Black Utahans took place at the Abraham Lincoln Colored Club. But at that 1898 meeting, Julius Taylor made a riveting presentation on Black rights, arguing that the 1896 Republican platform paid only lip service to the needs of the Black community. Since the GOP was not willing to put their money where their mouths were and actually stand up for a group of people who had shown them tremendous loyalty, Julius Taylor claimed they had a lot of nerve to use Black rights as a political issue. William

Taylor disagreed vehemently, and each publisher went on to use his newspaper as a platform to defend his position. The feud between the two newspapers ultimately precipitated the emergence of support among Blacks for the Democratic Party.

The solid support Republicans had historically received from Blacks was beginning to weaken. It was, however, the Republican William Taylor who was the first Black American to run for general elected office; he was nominated for a seat in the Utah House of Representatives in 1897. Though Taylor received fewer votes than any other candidate on the ballot, he still totaled an impressive 6,542 votes. This was stunning proof that, over a hundred years before Barack Obama and over sixty years until the civil rights movement, several thousand White Utahans were willing to cast their ballots for a Black candidate.

THE FIRST OF A LONG LINE

Around this time, America's first Black millionaire was building her fortune. Sarah Breedlove was born two days before Christmas 1867, in the sleepy village of Delta, Louisiana. Her early life was hard and by no means atypical for a young African American girl of that time. At the age of seven she lost both her parents to Yellow Fever Epidemic of 1873, and by the time she was ten years old, was working as a maid in Vicksburg, where she and her sister had moved to escape the epidemic. At the age of fourteen she married, not for love, but to escape life in her sister's house. After ten years of rocky marriage, Sarah divorced and relocated with her only child, her daughter Leila, to St. Louis, Missouri, where she worked as a laundress and started to attend night school.

In 1906, Sarah changed her luck. She invented a hair straightening process for African Americans. Using the name of her second husband, Charles Joseph Walker, Sarah began to market her invention under the name Madame C. J. Walker. Though her invention was an elaborate ritual involving brushes, combs, and heat, it was none-theless an immediate success. In 1908, she opened a training school for beauticians in Pittsburg—the Leila College for Walker Hair Culturists—that taught African American women how to become "Walker agents" and sell her hair care products door-to-door. By 1910, there were over a thousand Madame Walker agents across the United States, and by 1914 she had earned her first million dollars.

Madame Walker died in May 1919, at the age of fifty-one. She left behind a company that had expanded to include offices in Indianapolis and New York City, a thirty-four-room mansion in Irvington-on-Hudson, New York, and a legacy of inspiring achievement.

CHANGE OF AGES

In 1915, Booker T. Washington died, and with him his personal organization that linked Black leaders throughout the nation and enabled them to speak with one voice for the Black community. In his place came W. E. B. Du Bois, a relatively more outspoken leader of a northern group who rejected Washington's focus on sep-arate development for the Black community and demanded access to national political discourse—and recourse.

It was during these early years of the twentieth century, as the new populations of urban centers became more racially diverse—and, in tandem, progressively more segregated—that many former predomi-nantly White middle-class neighborhoods began transforming into fashionable neighborhoods for the cities' African American newcom-ers. Between 1915 and 1920, nearly one million African Americans moved to northern cities and found themselves "confined"—if not exactly by what the era euphemized as "mutual consent"—to the areas abandoned by Whites. There was Oak Hill in Grand Rapids, Iowa, a former home to affluent Whites that became a thriving working-class Black neighborhood. There was U Street in Washington, D.C., an avenue designed predominantly by Black architects and home to many prominent Black leaders such as Thurgood Marshall and Duke Ellington. It became known as the "Black Broadway" for the richness of the jazz clubs and legendary performers who appeared in the clubs along its glittering entertainment corridor. There was East Elmhurst, in Queens, which Blacks had called home since the time of the Great Migration; this is the neighborhood in which Eric and Miriam Holder settled when they immigrated from the West Indies in the 1950s, and where they raised their son Eric, now the United States' first African American Attorney General.

There was Beale Street in Memphis, Tennessee, which during Jim Crow was an oasis of Black business and social activity—and the center of the creation of the only truly native American art form, the blues. Black communities prospered from Montgomery, Alabama, to New Orleans, Louisiana, and even to Boston, Massachusetts, which despite having a thriving abolitionist community during the Civil War, had a relatively small Black population. Whole cities and towns such as Nicodemus in Kansas; Boley, Porter, and Langston City in Oklahoma; Booker and Kendleton in Texas; and Fairmount Heights; and Glenarden in Maryland were established as havens for Blacks in these early years of the twentieth century, becoming business, cultural, and intellectual sanctuaries for their residents.

Some of these sanctuaries, indeed, were summer resorts for the African American well-to-do. Martha's Vineyard, haven to prominent politicians from Kennedys to Obamas, has a long history as a seasonal resort for Blacks. Highland Beach in Anne Arundel County, Maryland is another such summer community. It was founded in 1893 by Charles Douglass, the son of the abolitionist Frederick Douglass, after he and his wife, Laura, had been turned away from a restaurant because of their race. Douglass bought a forty acre plot of land in Chesapeake Bay with 500 feet of beachfront and fashioned it into a popular seasonal gathering place for upper-class Blacks. Residents, and guests of the residents, have over the years included such luminaries as the entertainment legend Paul Robeson, the Washington, D.C. municipal court judge Robert Terrell, the poet Langston Hughes, the activist Harriet Tubman, author Alex Haley, tennis great Arthur Ashe, Bill Cosby, and W. E. B. Du Bois. While, sadly, Frederick Douglass himself was never a resident—he died before the house his son was building for him could be completed— and while the town is now less than 40 percent African American, many of the homes in Highland Beach are still owned and occupied by descendants of its original fathers and mothers.

DON'T BUY WHERE YOU CAN'T WORK

But just as people in the Black communities were becoming empowered under Du Bois' influence to shift from the confinement of "mutual consent" and take their place on the national stage, the Great Depression happened. The 1930s were an uncertain time for race relations in America. The growing presence of Blacks in northern cities increased tension between the races. The many New Deal programs that gave Black Americans opportunities that had so long eluded them in the past also put them in daily contact with and revealed their struggles to White Northerners.

> Such federal programs as The Federal Music Project, Federal Theatre Project, and Federal Writers Project enabled black artists to find work during the depression, often times creating art or stories which portrayed the historic and present situation of blacks in the South. Projects chronicling the lives of former slaves were also begun under the auspices of these programs. At the same time competition for Works Project Administration (WPA) jobs in the South during the thirties also brought to light the persistence of inequality even in the government. Since the WPA required that eligible employees not have refused any private sector jobs at the 'prevailing wage' for such jobs, African Americans (who were paid less on average than whites in the South) might be refused WPA jobs which whites were eligible for. Such discrimination often extended to Hispanic-Americans in the Southwest as well. Despite

such difficulties, WPA head Harry Hopkins worked with NAACP leaders to prevent discrimination wherever possible resulting in general support for programs (and the government) by the black community.[2]

Still, in spite of the jobs that were created through the programs of the WPA, Black unemployment rates were often double, even triple, the national average. The fuel of Du Bois' impassioned writing and oratory combined powerfully with the frustration sparked by lack of work, and the "Don't Buy Where You Can't Work" movement was born.

In 1929, the Chicago newspaper the *Whip* sponsored a boycott of Chicago stores that refused to hire Blacks. The boycott was supported by the newspaper's editor, Joseph Bibb, as well as the Reverend J. C. Austin of the Pilgrim Baptist Church. Their efforts resulted in more than two thousand Blacks in the city being hired, many of them as clerks in Chicago's department stores.

And, as can frequently happen with successful local movements, this one spread like wildfire to other cities. In New York, where the unemployment rate in Harlem was a staggering 50 percent, support for a similar boycott came from Harlem's newspapers, the Harlem Business Men's Club, and Black ministers, politicians, and other community leaders.[3] The Reverend John H. Johnson of Saint Martin's Protestant Episcopal Church formed the group Citizens League for Fair Play to organize picketing efforts. The legendary Adam Clayton Powell, Jr. was a proponent. In Washington, D.C., the New Negro Alliance, Inc., took up the motto "buy where you work—buy where you clerk." In response to local layoffs of Black workers, their picketers targeted stores within the city's Black district, such as the A & P, Kaufmans, and the High Ice Cream Company. Over time, working in common cause at their local levels, these groups formed comprehensive agendas for increased Black employment and opportunities—and a national network to promote them. Their legacy? As with every other incremental social and financial advancement we've discussed so far, these organizations fostered enlightenment and aspiration within the Black community. The protests of the 1930s helped create the first affirmative action hiring programs in history, as well as the model for direct-action civil rights protests, like lunch counter sit-ins and bus boycotts. The "Don't Buy Where You Can't Work" movement paved the way for the legislative progress that was the culmination of the 1960s civil rights movement.

THE CIVIL RIGHTS ERA

The period we know now as the civil rights era lasted just fourteen years. It spanned the years between 1954, when the United States Supreme Court handed down its landmark ruling in *Brown v. Board of Education* that school segregation was unconstitutional, and 1968, when Dr. Martin Luther King was assassinated the day after delivering his famous "mountaintop" sermon in Memphis, Tennessee.

In between these seminal events came Rosa Parks and the three hundred and eighty-one day Montgomery Bus Boycott, in which African Americans—and a few White supporters—walked to work, reducing bus revenue by 80 percent and thereby forcing the desegregation of public transportation in the city. There came the first student sit-in at a Woolworth counter in Greensboro, North Carolina, protesting the exclusion of Black patrons, and the Freedom Rides on interstate buses that ended segregation for passengers engaged in interstate travel. There came voter registration organizing across the southern states of the Union, the 1963 March on Washington, and the Mississippi Freedom Summer in 1964. There came the passage of the Civil Rights Act of 1964, a bill introduced by President John F. Kennedy and passed by Congress during the administration of President Lyndon Johnson, that outlawed racial segregation in schools, public places, and places of employment. Then came the Voting Rights Act of 1965—again signed into law by Johnson—that

suspended poll taxes and literacy tests, and authorized federal super-
vision of voter registration in states and in individual voting districts
where such tests were being used. There came the Nobel Peace Prize
for Dr. Martin Luther King at the age of thirty-five, making him
the youngest man ever to be so honored.

In other words, between the years 1955 and 1968, all those deci-
sive events that get so much ink in our history books occurred: the
hard-won victories of the parents—or *grandparents*—of the AAAs
we're talking about in this book. The civil rights era was a legendary
time, filled with heroes, but like the flying aces of World War I or
King Arthur's knights, our contemporary AAAs learn about those
days only in tales.

To understand how far removed AAAs are from this era, we can
look at the present challenges facing the flagship institution of the
civil rights movement, the National Association for the Advancement
of Colored People (NAACP). The NAACP faces "tough criticism
from activists who argue that it's grown out of touch with the
grassroots and has failed to deploy new technologies and tactics to
encourage broader participation. At the same time, the changing
politics of race in America have made the group's agenda less obvious and more difficult to articulate."[4]

Part of the problem, certainly, has been its own success. The obvious and well-articulated objectives of the 1950s and 1960s have been
well met. According to Melissa Harris-Lacewell, an associate professor of politics and African American studies at Princeton University,
"one obstacle the NAACP faces is in part due to the accomplishments of the previous generation of civil rights leaders."[5]

William Jelani Cobb, an associate professor of history at Spelman
College, defines the institution's struggle to maintain relevancy in
this way:

> We are more and more products of institutions that may not be
> African-American ones. When we look at people politically, like
> the Cory Bookers, the Barack Obamas and so on, these are folks

who are as comfortable in white environments as well as majority-black environments. I think the coalitions we work with will be very different.[6]

Harris-Lacewell looks confidently to the organization's new president, thirty-six-year-old Ben Jealous, appointed in 2008, for the necessary rejuvenation: "because he's part of our generation, who were born when many of those battles had been won..."[7]

The roots of the contemporary AAA generation were certainly planted in the civil rights era; the accomplishments of the time, as well as the remarkable leaders and venerable institutions that emerged from it, are honored. But with so many of the battles won, there was no place to go but up.

I SPY JULIA

It's an understatement to say that the generation of children who grew up in the 1960s and 1970s as part of the expanding Black middle- and upper-middle classes did so under the radar of the mass media. Because television is—or, at least *was* then—the reigning media, we can use the TV shows of the time as a sort of barometer of this phenomenon.

By 1956, in the years just prior to the beginnings of the civil rights movement, Nat King Cole was an international star, a top night-club performer with several million-copy hit records, including the unforgettable "Mona Lisa." He was a frequent and ratings-friendly guest star on the mainstream variety shows hosted by Perry Como, Milton Berle, Ed Sullivan, and other White entertainers. But he very much wanted a show of his own, and NBC agreed he should have one. The Nat King Cole Show began airing as a fifteen-minute weekly musical variety show in November, 1956—without commercial sponsorship. The network had decided to carry the bill for the show itself, in anticipation of attracting a national sponsor for a program of such musical excellence. But advertising agencies were unsuccessful at convincing clients to buy time on the show, "fearful that white Southern audiences would boycott their products."[8] The show was cancelled.

It wasn't until 1965—again with the support of NBC—that television got its first regular Black leading man. Bill Cosby paired with Robert Culp in *I Spy*, an adventure series in which the actors played globetrotting Pentagon spies. Audiences ate up the slick and snazzy *I Spy* for nearly four years, and the show's success helped pave the way for more Black actors in guest roles on other regular television series, including the popular *Dick Van Dyke Show*. But the portrayal of African Americans on TV as debonair international spies or typical middle-class suburbanites was certainly not the norm. From a virtual absence of people of color on television in the 1940s through the 1960s, the 1970s were a period of what the writer Christine Acham calls "hyper blackness."[9] The shows of this era were roundly criticized as being too rosy in their portrayals of racial relations (Diahann Carroll, the star of *Julia,* was a nurse and Vietnam War widow who seemed to live in a bubble of racial harmony), distastefully stereotypical (Flip Wilson's character, the sharp and stylish Geraldine Jones, the originator of the phrase, "What you see is what you get," as a throwback to Sapphire in *Amos 'n' Andy*), or frankly lame for their lack of any real social commentary (Wilson, for example, who avoided social critique, was on the air for four years, while Richard Pryor, who was far more bitingly critical in his social commentary, had a show that lasted for four *episodes*).

Tellingly, the Black viewership of these shows divided down not a racial line, but a class one: working-class Blacks appreciated the use of traditional Black humor and entertainment techniques and tuned in, while middle-class and upper-middle-class Blacks—the parents and grandparents of today's AAAs—tuned decidedly out. These shows contained little that related to their own *class* experience, or to the class experience of their children.

ADVERTISING TO THE BLACK AUDIENCE

Television, of course, wasn't the only media to fail in representing the realities of Black middle-class life, converging around stereotypes to dismiss a whole segment of high-achieving African American families. A look back through advertising archives reveals nearly a hundred years of marketers missing the mark. In fact, looking back on the advertisements that featured images of Blacks in the early part of the twentieth century is not just offensive—it is downright cringe-worthy.

Though ads featuring Blacks in the 1960s increased in number, "the manner in which they were portrayed tended to confirm and perpetuate racial stereotypes." Later, however, "improvements" in the occupational status of Black models in magazine ads were evident. There was an increase in the use of Black models between 1967 and 1974, but they "tended to advertise personal items, such as hair products, rather than nonpersonal products, such as automobiles."[10]

However, the 1960s and 1970s did bring some small breakthroughs in the way Blacks were represented in media. The young and growing generation of middle-class Blacks was taken seriously by advertisers like never before. Levi's, Lee, and J.C. Penney competed to dress the rising Black middle class. Greyhound appealed to Black travelers with an ad featuring two napping children riding together side-by-side—a young, White boy dressed as an Indian and a Black boy dressed as a cowboy. Atlantic City advertised its "boardwalk bonanza" with print ads showcasing a young Black couple building sandcastles on its beach. Armstrong Tires targeted this audience with a rather strange ad: a photo of a young Black woman posed seductively beside a tire under the headline "the tire for lovers."

Not all advertising to this audience was a step in the right direction, of course. Johnnie Walker Red, Hiram Walker bourbon, and Martini and Rossi vermouth were among the legal beverage ads that featured Black faces enjoying their products, as did ads for Viceroy, Tareyton, and Kent cigarettes—and it is the ads for these kinds of adult products around which controversy still swirls. In 1991 a report from the Stanford Center for Research in Disease Prevention did an analysis of billboard placement in neighborhoods in San Francisco, California. What they found was that "across all billboard advertising of products and services, tobacco (19%) and alcohol (17%) were most heavily advertised." Furthermore, "Black neighborhoods had the highest rate of billboards per 1000 population," and "Black neighborhoods were proportionately more likely than other neighborhoods to have billboards advertising menthol cigarettes and malt liquor."[11]

Twelve years later, these industries were still disproportionately targeting the Black audience. A 2003 study by Georgetown University's Center on Alcohol Marketing and Youth found that, of the $333 million in advertising placed by the nation's alcohol industry that year, "Blacks from 12 to 20 years old saw 77 percent more of these ads...than their non-Black peers did."[12]

It was obvious the industry was "directly targeting Black kids," according to the Reverend Jesse Brown, executive director of the National Association of African Americans for Positive Imagery. "African American kids tend to be trendsetters in what they buy, so the industry thinks if it can get more African American kids to buy, it can also get their White counterparts to buy."[13]

It's not merely inappropriate placement, or overplacement, of ads that expose young Blacks to adult or unhealthy products that remains a disappointingly contemporary issue. But let us suggest that some of the lingering problems with the limited or misrepresentative images of African Americans in creative has to do with the historic lack of Blacks in the executive offices of advertising agencies. Through the decades, though Blacks were targeted by advertisers, and there were some Black faces in print and television ads, the people who created those ads were rarely themselves Black. In fact, there was a time when it was so rare to find an African American in an executive position in an advertising agency that the 1970 movie *Putney Swope*—tagged as "the truth and soul movie"—featured a comedic storyline about an advertising firm that *accidentally* voted in the Black partner as the new head of the company.

In real life, of course, things weren't quite that comical. The late Vince Cullers of Chicago launched the first Black advertising agency in 1956, while Luis Díaz Albertini founded the first Latino shop, Spanish Advertising and Marketing Services, in 1962 to attempt to both appeal to these minority segments as well as improve the images that represented them in the mass media. Jason Chambers, a professor of advertising at the University of Illinois, tells the inside story of the history of Blacks in the ad business in his book, *Madison Avenue and the Color Line: African Americans in the Advertising Industry*. He writes that Blacks in advertising in the early days of the 1960s and 1970s saw themselves as having a "dual responsibility" to their clients to sell products to the Black community while working to change its often negative image.[14] "Their lowly status in advertisements confirmed their economic disenfranchisement, just as violence and Jim Crow laws confirmed their political disenfranchisement," Chambers wrote, but "getting to that point...first required getting white-owned companies to recognize the black consumer market."[15]

Getting companies to recognize the underserved and generally untapped Black market—especially the AAA market—is our on-going challenge. And companies like Lagrant Communications, Muse Communications, and UniWorld help marketers reach multicultural consumers. These are important moves forward as we estimate that of total ad spending in the United States, less than 4 percent is directed at any level of minority consumer.

Still, for every Allstate ad featuring a Dennis Haysbert (the actor perhaps best known for portraying America's first Black president on the television show *24*), there are two or more that feature

African Americans in stereotypical (i.e., dancing or playing sports) or even subservient roles. In one strikingly unfortunate image, Intel advertised its Core2 Duo desktop processor with a photo of six athletic Black men in cubicles bowing down to one White man who has arms cockily crossed and a smug smile on his face. This image is under the headline, "Multiply computing performance and maximize the power of your employees." We wonder if Intel gave a thought to the evocation of subservience in this ad—or to how that image might play with African American consumers.

It is ads like this one that can make us appreciate Gucci's marketing campaign with Rihanna. However misguided these ads are as an attempt to attract the AAA segment, they do feature the image of a beautiful, strong, and empowered Black woman.

JACK AND JILL

So advertising has a way to go to fully understand the subtleties intrinsic in marketing to the African American audience and, especially, the AAA segment. And those old television shows with their mostly or all Black casts, rerunning seemingly endlessly on cable channels, neither reflect the background of the AAA group, nor resonate with their experience of growing up. But what was a part of the childhood experience of many of today's AAAs, and was a key component to the foundation of many of their lifestyle choices, was a group called Jack and Jill.

Jack and Jill of America, Inc., founded in 1938 in Philadelphia, Pennsylvania, was the brainchild of Louise Truitt Jackson Dench, a mother who wanted Christmas to last all year long. "I went to Brooklyn once to visit a friend," Dench once said. "She was telling me what a wonderful Christmas they had had with all their visiting friends who had moved away into various boroughs and cities. They had all these children now and everyone came back and had a Christmas party. I enjoyed the story and I said, 'Gee, Philadelphia could enjoy something like that, but I want our group to be a club permanently, not just get-togethers at Christmas.'"[16]

It was Marion Turner Stubbs Thomas who ran with Dench's idea, and founded Jack and Jill. She shaped it as a family organization for middle-class African American children, providing cultural, social, civic, and recreation services. In her words, the organization was a "means of furthering an inherent and natural desire... to bestow upon our children all the opportunities possible for a normal and graceful approach to a beautiful adulthood."[17] In other words, Jack and Jill was intended to give Black children the entrée, through its neighborhood programs and schools, to a privileged lifestyle with like-minded peers just as their White counterparts experienced in youth organizations, such as the Junior League or the DAR, that were, again by law or custom, restricted to a segregated membership. Jack and Jill created an environment and a set of expectations—and early links to an aspirational social network—that helped put African American kids on the road to affluence. The group's mantra is that any child, given the proper opportunity and guidance, can be a leader. The group's aim is to improve the quality of the children's lives, and help them enter adulthood with a sense of sophistication and self-worth that will serve them well in their goal of professional success.

As Len says:

> I am certainly a product of this generation of upper middle class parents. You know, I didn't see myself as rich but we didn't want for anything. My father was an entrepreneur. My mother didn't have to work. I was in Jack and Jill. I went to a private Catholic high school. I went to both a "Big Ten" school and a historically Black college. At this point in my life, I've run several ground-breaking businesses and my wife and I are busy making sure our kids are ready for—you guessed it—Jack and Jill.

Jack and Jill of America, Inc., is the oldest and largest non-profit African American family organization in the United States. It remains one of the most important resources Black parents look to in charting the future prospects of their children. It boasts a membership of over 30,000 children, from the ages of two to nineteen, and their parents—who benefit from the organization's network as much as their offspring do. Lawrence Otis Graham relates this story about Jack and Jill in his book *Our Kind of People*:

> "Well, what I'm really hoping is that she'll become a litigator or, maybe some kind of judge," uttered a woman, with total confidence, as we looked out over Central Park from a Fifth Avenue apartment in the East Seventies. Dressed in a conservative linen suite, the forty-three year old mother adjusted her Chanel scarf in the reflection of the large living room window...Looking across the room at twelve year old Laura, I shrugged my shoulders...This was exactly why we had all been enrolled in Jack and Jill as kids.[18]

Today Jack and Jill is more of a social organization, though its more profound founding purpose has certainly not been lost in a whirl of parties. Inspired by President Obama's request for all Americans to be agents of change by doing their part in improving their local communities, in 2009 the organization launched its Teen Conference Day of Service. The Day of Service events Jack and Jill sponsors are geared toward teaching young Black adults about the morals and values of volunteering, and in doing so nurture both the strong public service tradition within the African American community, and the next generation of philanthropists.

A WEALTH OF EDUCATION

As we noted earlier in this chapter, esteem for education in the Black middle class is rooted in the deep respect for Black teachers that was a part of the community from its founding days. Like most ethnic and immigrant parents, Black parents have historically understood that the road to a better life for their children is paved with degrees from institutions of higher learning. However, for the alumni of

Jack and Jill—the kids of middle- and upper-middle-class parents of the 1960s and 1970s—getting a good education wasn't anything out of the ordinary. Going to college was no longer an aspiration within this segment, it was a given. What was aspirational was the level of success they could achieve after college, and in many cases grad school, was behind them.

Some AAAs spent their college years at a historically Black college. There are 103 such colleges today in the United States, and that includes public and private, two- and four-year institutions, from community colleges to medical schools. These were often the schools that were attended by the first children in Black families to get college degrees, and constituted for a generation of Black families the primary route out of poverty and into professional careers. While schools such as Howard University and Morehouse remain some of the finest teaching institutions in the country, many of today's AAA professionals have attended non-Black colleges as well. Barack and Michelle Obama at Harvard, Eric Holder at Columbia, or Desiree Rogers at both Wellesley *and* Harvard might immediately leap to your mind as examples. Or perhaps you think first of entertainers and sports figures such as Oprah Winfrey (Tennessee State University) and Michael Jordan (University of North Carolina at Chapel Hill). But Black alumni groups from schools such as the Wharton School of Business, MIT, and other Ivy League schools, as well as Black professional organizations such as the National Black MBA Association, the National Society of Black Engineers, and the Executive Leadership Council, put the lie to the assumptions inherent in such a small sampling. To broaden your view, go visit the Web site of an organization called Black Excel.

At www.blackexcel.org, Black Excel, a college guide recently cited by both *Ebony* and *Black Enterprise* as a major resource for young African Americans and their families in navigating the college admissions process, maintains a list of contemporary Black leaders whose names you might not recognize, but who graduated from colleges and universities all across America. Black Excel seeks to break the myth that higher achievement for African Americans is limited to a career in entertainment or sports: "Many of our 'stars' are unsung. An extended list of their success stories in academia, science, business and beyond could fill a library." Even a quick glance through this list will broaden your reference point for Black achievement. We'll give you just a taste here of some of the more well known, even iconic, figures who made the list, and you can go to the site for the whole feast (an asterisk denotes a historic Black college): Alcorn State University*, author Alex Haley and slain civil rights leader Medgar Evers; Amherst College, Charles Drew, the scientist who discovered blood plasma; Antioch College, Coretta Scott King and Congresswoman Eleanor Holmes Norton; Arizona State University, Reggie Jackson, baseball Hall of Famer; and Ron Brown, United States Secretary of Commerce—and that's just a quick sampling of the A's.

Many of today's AAAs are what the *New York Times* calls the "children of 1969"—the year "America's most prestigious universities began aggressively recruiting blacks and Latinos to their nearly all-white campuses."[19] The shift in policies of Ivy League colleges, as well as other celebrated institutions of higher learning to ones of diversity and inclusion followed close on the heels of the civil rights era. Nicholas Lemann, dean of Columbia University's Graduate School of Journalism and author of *The Big Test*, a history of the SAT and the rise of America's meritocracy, cites the shift as a crucial one in allowing "the children of 1969 to flow more easily between the world which their skin color bequeathed them and the world which their college degree opened up for them."[20]

In the next chapter, "Meet the Royaltons," we'll introduce you to some contemporary AAAs—people who are representative of this elite group who walk easily these days in the world at large. The diverse achievements of this group, and the interesting careers and avocations this segment pursues remain, however, and for the most part, under the radar of the mass media. In the meantime, here are some overall facts and figures for you to file away for reference: while the number of White students who enrolled in college increased by 3.4 percent in the years between 1993 and 2003, minority enrollments increased by 50.7 percent, with Black enrollment increasing by 42.7 percent.[21] Growth in this area is slow, but proportionately these are astronomical percentages. Comparatively, when you realize that in 1960, just a few years after *Brown v. the Board of Education,* when many of the parents and grandparents of contemporary AAAs were starting to pursue their own educations, the median years of school completed by Black men was just 7.7, and by Black women just 8.6, these percentages are through the roof, and growing.[22]

THE INVISIBLE PEOPLE

Why are so many AAAs under the radar? Why does it involve so much research to develop the marketing strategies that will make your brand relevant and significant to this target audience? As we've already mentioned, focusing solely on Black athletes, celebrities, and rappers could land you way off the mark. These one-dimensional references—hip-hop and urban—have overshadowed the very real "silent majority" that is this affluent group.

Part of the reason for the overshadowing lies, as we've pointed out, in misrepresentation by the media. Soledad O'Brien hit the mark succinctly in her CNN special, "Black in America," when she pointed out the "incomplete picture Americans have of black America; incomplete because many success stories are missing." The psychiatrist Dr. Carlotta Miles agreed with O'Brien when the interviewer posed this question: "Do you think most Americans, whether they're black Americans, white Americans, any American has no clue that privileged, wealthy, well-connected black people exist in decent-sized numbers?" Miles answers quietly, labeling the AAA group to which she belongs "the invisible people."[23]

But it's also instructive to know that part of the reason today's affluent African Americans are less conspicuous is that they intend to be.

For AAAs, living the good life—living in luxury—is not something that's unusual. And, because it's not unusual, it's not something
to brag about. You will never see an AAA appearing on, for example,
the likes of MTV's show *Cribs* (or, for that matter, even watching it).
They have no need to promote their ability to afford the lifestyle they
lead because it is a lifestyle they have grown up *expecting* to lead.

Unlike the Vanderbilts, Rockefellers, Carnegies, and the Mellons
who flaunted their fortunes around the turn of the nineteenth century, AAAs are characteristically low key in acquiring and appre-
ciating the finer things in life and in keeping those fine things in
the family. So, if over-the-top mansions and flashy cars are out, let's
look at what are the real core concerns of wealthy Black families.

According to a report from Northern Trust[24] DreamMakers
Forum in 2008, approximately half of affluent Blacks feel that providing financial support to family members is a chief responsibility,
with six out of ten providing support to their parents, and two in
five providing support for siblings. The most important need affluent Blacks believe they will meet in providing this support to adult
family members is long-term or disability care. More than three in
five affluent Blacks say they are either extremely or very concerned
about ensuring that the next generation of their family members
lead productive, meaningful lives amid continued affluence, and
more than half believe it is either extremely or very important to
leave an estate to their heirs. More than half have a will in place—in
contrast to the general population in which only 42 percent have a
valid, legal will.[25] It is also common for affluent Blacks to name a
family member the executor to their estate. In research conducted
by Diversity Affluence in 2008, AAAs were commonly concerned
about leaving a legacy and wealth creation. Passing on financial values and effective strategies for dealing with wealth is important to
this segment, and establishing trusts for estate and tax planning

purposes is common, and especially at the top of the to-do lists of affluent Gen X and Y Blacks. In terms of charitable giving, AAAs give to their chosen philanthropies on a grand scale. But giving in this segment is focused on domestic and, often, local organizations (as opposed to large national and institutional charities; the number one goal that affluent Blacks hope to accomplish with their giving is to support a cause in which they personally believe). Additionally, the majority of this group believes that it is more important to give during their lifetimes, rather than through a will and, on average, affluent Black individuals donate upwards of $35,400 to religious organizations or charitable causes annually.

The core concerns of this group, in other words, are possibly more aligned with what we think of as traditional, conservative financial goals than you might have imagined. Marketers who have a grasp of this fact—especially those whose clients are in the business of investment advising—are going to be advantaged in understanding and meeting the needs to the AAA consumer.

From Entrepreneurs to Everyday People

Black wealth in America is largely a twentieth-century phenomenon. But like their White counterparts—the Vanderbilts, Rockefellers, Carnegies, and Mellons we mentioned earlier—AAAs are truly the builders of what is a whole new circle of influence and power. At the center of this circle are a generation of pioneers; people such as Ben Carson, Bob Johnson, Don Peebles, and the late Reginald Lewis, storied entrepreneur and CEO of Beatrice Foods, Inc.

Lewis, born into the middle class in Baltimore, Maryland, in 1942, attended what is now Virginia State University on a football scholarship and went on to graduate from Harvard Law School in 1968. He was recruited to the top New York law firm Paul, Weiss, Rifkind, Wharton and Garrison, LLP, immediately following law school, but left after only two years to start his own firm. In 1983, after fifteen years of practicing corporate law, Lewis created the TLC Group, L.P., a venture capital firm. In his first major deal he purchased the McCall Pattern Company, a home-sewing pattern business, for $22.5 million. Fewer and fewer people were sewing at home and the McCall Pattern Company was seemingly on the decline. While it had posted profits of $6 million in 1983, it was a risky move for Lewis. But he negotiated the company's asking price down, and raised over one million of it himself from family and friends and, within just one year, he'd turned the company around.

How did he manage to do what had appeared to be an impossible task? Through thinking that was so out of the box for the time we have to think that the box wasn't even in his peripheral vision. He found a new use for the company's machinery during downtime, manufacturing greeting cards, recruited top managers from rival companies, put an emphasis on new product development all while containing costs. Then he began to export to China. This strategy led to the company's most profitable year in its history. When he sold the company it was at a profit that gave his investors a 90:1 return—and Lewis himself had held over 80 percent of the company's stock.

In 1987, Lewis bought Beatrice International Foods, a snack food, beverage and grocery store conglomerate, from Beatrice Companies for $985 million. At the time, it was the largest African American owned and managed business in the United States, and it became the first Black owned business to have more than $1 billion in annual sales.

Before dying in 1993 from brain cancer at the young age of fifty, Lewis was listed among the 400 richest Americans with a net worth estimated at $400 million in the *Forbes* magazine.

"It's understandable that [my race] is something people focus on, but what I focus on and what others focus on are two different things. I focus on doing a first-rate job on a consistent basis...I would say my race hasn't been a factor one way or the other," Lewis wrote in his autobiography *Why Should White Guys Have All the Fun?*[26]

Lewis's intense focus on consistent quality would appear to be a legacy to the business men and women of color who followed him. The most recent United States Census data shows that African American owned businesses are growing at a faster rate than all other minority-owned businesses, and that minority-owned businesses are growing at a rate that is greater than the overall national average. This rapid growth is attributable to the rise of Black urban professionals in the corporate world in the 1970s and 1980s, and the subsequent rise of urban/hip-hop culture in the 1990s that spawned a whole host of Black businesses related not only to urban/hip-hop music but to such offshoots of its culture as movies and fashion. These businesses benefited not only the people who owned the companies but all the people who worked for them as well. According to the Minority Business Development Agency, in 2006 there were 1.2 million African American owned businesses that were generating $89 billion in revenues—an increase of 56 percent since 1997.

And interestingly, businesses owned by women of color are a substantial economic force. The Center for Women's Business Research's biennial update of the number, revenues, and employment trends for these businesses was released at a symposium. The findings, underwritten by Wells Fargo, showed strong growth. Between 2002 and 2008, the number of such firms increased by 32 percent, their revenues by a dramatic 48 percent, and their employment by 27 percent. The Center estimates that as of 2008 there are 1.9 million firms owned by women of color, employing 1.2 million workers and bringing in $165 billion in revenues. Firms owned by women of color comprise 26 percent of all women-owned firms. A prestigious consortium of corporate and association sponsors funded Accelerating the Growth and Success of Businesses Owned by Women of Color. Over the time frame of the study, a total of fifteen corporations invested in the research. In 2008, the corporate sponsors included Wachovia Corporation; Wells Fargo Bank; Verizon; OPEN from American Express; Time Warner; UPS; Capital One; MasterCard; American Airlines; Ernst & Young; United States Postal Service (USPS); and PepsiCo.

GATHERING THE ROCKS

The story of Reginald Lewis, and what he was able to accomplish in his short life, is an undeniably inspirational one to members of minority populations with similar aspirations. Much in the way that the stories of Meg Whitman or Carly Fiorina motivate women, Lewis has more than earned his place in the pantheon. Moreover, we can easily say that Lewis is held in esteem not only by Black businessmen, but that such were his innovations and accomplishments that he is held as a role model by all striving businesspeople regardless of race.

But where the truly revolutionary nature of Lewis's success reveals itself is in the place it intersects with the accomplishments of another Black leader of the time, Jesse Jackson. In the early- to mid-1980s, when Lewis was building his fortune, Jackson staged two presidential campaigns. It was during Jackson's 1984 announcement in Philadelphia that he was going to run for the Democratic nomination that he gave what many remember as his famous "rocks" speech.

In this speech, Jackson inveighed that Reagan had won the 1980 presidential election by default, while the majority of the voters were "asleep":

> I close with another story of a little shepherd boy named David. Everybody in town was scared of Goliath. [But] little David took what God gave him, a sling shot and . . . a rock . . . What we need, we need to organize . . . and win because we are going to stop the rocks that's been lying around and pick them up. In 1980, Reagan won

Massachusetts by 2,500 votes! [But in Massachusetts] there were over a hundred thousand student unregistered, over 50,000 blacks, over 50,000 Hispanics. He won by 2,500...He won Illinois by 300,000 votes—800,000 unregistered blacks, 500,000 Hispanics, rocks just laying around...In 1980 Reagan won eight southern states by 182,000 votes when there were three million unregistered blacks in those same eight states. Rocks just laying around! He won New York by 165,000 votes. Six hundred thousand students unregistered, 900,000 blacks, 600,000 Hispanics. Rocks just laying around...

We all know that Jackson's 1984 and 1988 bids for the Democratic nomination were unsuccessful. We acknowledge that 2008 gave us a different kind of African American candidate. That difference is perhaps best explained by the scholar Henry Louis Gates, Jr. who, in commenting on Obama's tenure at Harvard in contrast to Jackson's at the HBCU North Carolina Agricultural and Technical State University, said, "It would have been impossible for Barack Obama to go from a historic black school to become president, at this time. The whole point is that a broad swath of American had to be able to identify with him."[27] But what should also be remembered is that Jackson's call to organize and register the votes that had lain around like so many impotent rocks in 1980 was prophetic—exactly a part of what happened to bring Obama's victory some twenty-eight years down the road.

Even more critically, this call to action was not interpreted as a call only to political organization. Jackson's losses took place against the backdrop of increasing successes for Black business leaders and against stunning successes on the part of leaders like Reginald Lewis. The call resonated—if not overtly, or even always consciously—as one to organize *financially*. An increase in wealth is exactly what happened. "Without wealth, groups are at a disadvantage for influencing public policy. Political contributions play a role in setting the economic and social agendas at every level of government. Wealth is also a means of social power because of the status associated with wealth."[28] For our purposes here we are not defining wealth as the sort of "super" wealth that brings to mind a Warren Buffet or a Reginald Lewis, but as the annual income and equity in homes, cars, savings accounts, stocks and other investments that allow for the comfortable upper-middle-class lifestyle contemporarily enjoyed by AAAs—that is, a minimum annual income of $75,000 for single people, and a minimum annual income of $150,000 for families.

THE "TYPICAL" AAA

Let's revisit some key points and sum up all we have learned about the AAA audience in this chapter.

First, and obviously, AAAs are wealthy, and their wealth was earned by way of a rigorous education combined with Reginald Lewis-like hard work and focus on consistent quality. Most AAAs are not sports celebrities, and their only likely relationship to hip-hop was enjoying the music along with their peers when they were in their twenties. While AAAs still enjoy listening to hip-hop on occasion, the selection of tunes on their iPods has become diverse as their tastes have matured. They have developed an affinity for luxury—such items as classic clothing and high-performance cars, fine art and wines, and not just as consumers but as collectors. Their annual incomes, as well as the investment know-how their parents gave them while they were growing up, have enabled them to own a luxurious home or two, employ a personal trainer, indulge preferences for high-end jewelry and accessories, travel extensively and, generally, to enjoy an enviable lifestyle. These lifestyle amenities are not novel—AAAs are not *new* consumers but part of a network of that stretches both wide in the present day, as well as deep into the past. Though it may seem that way to the untrained eye, like most of their contemporaries, AAAs live under the radar—partly because the mainstream media has not generally taken notice and reported the story of affluent African Americans to the population at large, and partly because AAAs are understated and consume with discretion.

Like the majority of their peers, most AAAs also donate a portion of not only their money but their time to good causes. Many sit on the boards or advisory committees of philanthropic and cultural organizations, particularly those organizations that have cultural relevance such as Evidence, A Dance Company. The mission of the Evidence Dance Company, as stated on their Web site, www. evidencedance.com, is "to promote understanding of the human experience in the African Diaspora through dance and storytelling and to provide sensory connections to history and tradition through music, movement, and spoken word, leading deeper into issues of spirituality, community responsibility and liberation." An AAA often finds the work that he or she does for such a cultural organization personally fulfilling for several reasons—it is a way to celebrate and honor their ethnic heritage, as well as to support a city cultural institution. It also keeps them connected to a network of other like-minded people who support the organization by attending its performances and galas, and/or through corporate partnerships.

So what is it that you would like to sell to such a wealthy, well-connected consumer? A new winery in Sonoma County, California, that has just bottled its first vintage of Rhone-style syrahs, or an old one in Napa Valley that wants to expand its customer base? A vacation in the south of France? An Hermes tote or a Ferragamo suit or luxury car or boat?

How do you get your product and services on their radar?

How do you get them to say nice things about your product when their friends admire the purchase?

How do you solidify your brand as part of their lifestyle?

- Shift with the new demographic paradigm. Recognize that you want—and that you need—the AAA segment.

- Assess the opportunities for marketing to them that are measurable, scalable, and very likely waiting right outside your door for you to take advantage.
- Adapt your marketing mix to appeal to these influential buyers—or you might be leaving yourself open to the competition who will.
- Learn how to click with this new niche in a culturally competent, 360 degree way. Know who you are marketing to.

Come and meet the AAA Royaltons.

CHAPTER 3

MEET THE ROYALTONS*

In 2006 we were looking for a succinct way to describe the people who made up the emerging affluent ethnic segment of the market. To describe these consumers as "minorities" seemed inaccurate. After all, the niche markets—the wealthy African Americans, Hispanics, and Asians that Andrea's company, Diversity Affluence, helps marketers to reach—were quickly becoming the collective majority.

We also wanted to move away from the stereotypes often associated with the word "minority." The pigeonholes "minorities" supposedly occupy sit in jarring contrast to the reality of the total number of affluent ethnic households in the United States. Now estimated at over 1.3 million, these households have a total income approximately $387 billion, and a purchasing power exceeding $282 billion. Between lack of information and research, the lingering stereotypes associated with the word "minority," it was no wonder that marketers were virtually ignoring this consumer group. We wanted a precise word to point an accurate spotlight on the breadth and depth of this powerful segment.

So let's put the AAAs—the African American Royaltons—into the appropriate context. In order to do this, it is important to first clarify the definition of affluence. The research that Diversity Affluence has undertaken over the last few years has brought us to the realization that different organizations define affluence

---

Royalton (roi-el-ten) n. An affluent ethnic consumer of any non-White ethnicity.

differently. Some say that it's $75,000 per household, while others such as IpsosMendelsohn define it as $100,000. It's important, therefore, to note that all of the research Diversity Affluence conducts generally defines a minimum annual salary of $75,000 per individual (or $150,000 per household) as the baseline of affluence. Still, according to Bob Shullman, president of IpsosMendelsohn, 20 percent of households in the United States are affluent. Diversity Affluence's research from the January 2009 Multicultural Wealth Report shows that Royaltons account for 17 percent of the total affluent population, and that African American Royaltons account for a little over 32 percent of all Royaltons.

A 2008 report from Packaged Facts states that "There are 2.4 million affluent African-American households with annual incomes of $75,000 or more in the United States. They account for 17 percent of all African-American households but hold 45 percent of total African-American buying power."

You can see that there are varying figures and facts used to identify this segment, but the bottom line is that by anyone's definition, AAAs have the resources to be a market force. Research by Diversity Affluence shows that in the mid-Atlantic region alone (New York, New Jersey, and Pennsylvania), there are over 147,100 individual AAA Royaltons over the age of sixteen who earn $175,000 or more annually, and 69,800 households that earn $150,000 or more. Whether you are an executive ready to take your business to a higher level, a marketer who wants to reach new audiences in bold, new ways, a non-profit organization seeking to increase your donations, or a business development professional looking for new prospects and alliances, it's time to acknowledge that there's a large and dynamic group that's changing how America consumes luxury today.

Yet few marketers of luxury brands have Royaltons on their radar screens, and as a result have failed to capture a meaningful portion of expendable cash from AAAs. As stated in the previous chapter, it's important for marketers to know their audience in order to reach them on a serious scale. Traditionally, marketers rely on statistics, surveys, and focus groups to get to know their audience on a more intimate level. But if a brand is reaching out for the first time to a new consumer, how does it get to know this target audience better? To create a sophisticated and cost-effective strategy that goes beyond an ad or two, marketers must get to know Royaltons on a deeper level. What do they want? How do they consume? What is the best way for marketers to reach them? How can you adapt to and benefit from knowing them well?

Although the African American Royaltons represent a broad range of professions and careers, they share universal traits, interests, and desires that make them an understandable, distinguishable, and powerful segment of the adult U.S. population. They have the money, and they can't wait until smart marketers help them decide where and how to spend it. They represent an untapped, uncluttered market space for any company or non-profit organization to stand out in and capture—especially luxury brands such as BMW, Ralph Lauren, Apple, Chanel, Neiman Marcus, and others.

As Carl Brooks, president and CEO of the Executive Leadership Council recently told Andrea:

> There's an emerging group of African Americans that have achieved significant transferable wealth in this country. They have earned this through a myriad of avenues including real estate investment, entrepreneurship, and senior corporate executive leadership, among others.
>
> While we have traditionally targeted those in the professional sports and entertainment industries as the symbols of wealth in African American communities, there is a much broader group that deserves attention from premium luxury brands. It is perilous to fail to recognize the purchasing power and brand identification of this segment. They live right, they look right, they drive right, they vacation right, and have the ability to procure anything they want.

THREE DISTINCT SEGMENTS OF THE AFRICAN AMERICAN AUDIENCE

As we delve into the key characteristics that identify the African American Royalton, we think it will be helpful to start out by clearly distinguishing them from the general African American segment, as well as from the group with whom they are frequently confused, the urban/hip-hop crowd. As we've already pointed out, a percentage of AAAs may have been part of this youth culture ten or even fifteen years ago, but since then their tastes and purchasing power have matured and they have entered into a wider realm of aspiration. The following bullet list is a quick reference to differentiate the three groups and the nuances of marketing to each of them.

Urban/hip-hop:

- Average age 18- to 34-year-old young adults
- Ethnically diverse: Black, Latino, Asian, White
- Musical tastes focus on rap, hip-hop, R&B, reggae
- Attracted to the African American aesthetic
- Are a market driven by urban culture
- Have an appreciation for each of the broad range of cultures that make up the total demographic

African Americans:

- Cultural
- Average age 25- to 54-year-old adults
- African American
- Musical tastes: R&B, rap, gospel, jazz
- Traditional, loyal consumers
- Deeply entrenched in African American traditions, perspective, and way of life

Affluent African Americans:

- Average age 25- to 44-year-old adults
- African American

- College educated and/or entrepreneurially successful
- Broad range of musical tastes: R&B, jazz, classic hip-hop, gospel
- Maintain a keen sense of their heritage but are not constrained by it
- Have access to, consume, and desire the luxurious lifestyle aspects of all cultures

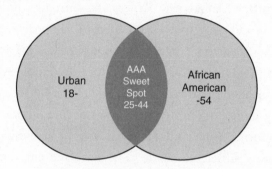

Make no mistake about it: these are three very different target markets. Understanding who AAAs *are not* is as critical in targeting this market segment as is understanding who they *are*.

INSIGHT INTO SOCIOECONOMIC AND RETAIL ROLES

Perhaps one of the most common misperceptions about the AAA segment, beyond that they are somehow interchangeable with a much younger demographic, is how they accumulate and relate to their wealth, so let us point out two striking facts.

First, AAAs are proportionately wealthier than their contemporaries in other segments. According to findings from Phoenix Cultural Access Group in 2008[1], AAAs hold an average of approximately $1.3 million in investable assets, compared to $992,000 in investable assets of the overall market. In addition, they show higher household incomes and slightly higher total net worth and total assets at an average of eight years younger (age 46) than their majority market counterparts (age 54). Their age has actually declined in recent years (ages 49–46), indicating opportunities to tap into a powerful and growing market that is far less likely to include retirees.

According to the same study, AAAs identify more with brand status than their peers in other niche market segments. A full 82 percent of influential African Americans—usually the more affluent of this group– say that brand drives their purchase decisions. This is in contrast to 68 percent of other U.S. "influentials" who say the same thing. But this attribute can be a double-edged sword. If the brand doesn't maintain a strong tie to the AAA segment, they will quickly abandon it for the next new hot thing. For example, the AAA community once embraced and bought in significantly high numbers the Cadillac Escalade. The Escalade enjoyed a remarkable 30 percent market share in the mid-1990s, but the brand did not take advantage of and embrace the audience through a strategic outbound plan. Along came the Lincoln Navigator, a brand that *did* embrace the audience through advertising in targeted media, included African American celebrities in their creative, and staged events focused toward the African American community. The result was that Cadillac lost a significant audience share in direct correlation to the share that the Navigator won and has been able to maintain ever since.

Byron Lewis, multicultural communications pioneer and president and CEO of UniWorld Group, the industry's longest-standing Black-owned multicultural advertising and marketing agency recently told Andrea: "Blacks over-index in a number of product categories, including automobiles, and this group of consumers will pay a premium for preferred brands."

Ford has been a UniWorld client for almost twenty-five years. Lewis was aware that African Americans represented the driving force in the nation's popular culture: music, sports, entertainment, and fashion. Having worked on Ford Motor Company's former

Land Rover and Jaguar vehicles, he knew that African American consumers were greatly influenced by status and style. The Lincoln Navigator's bold and dramatic design embodied these important characteristics and quickly became the SUV of choice in Black Hollywood. The trendsetting influence of Black athletic, hip-hop, and entertainment icons generated tremendous interest and sales among African American and general consumer audiences alike. Lewis says, "Even Suzanne DePasse, trailblazing producer of Emmy-nominated *Lonesome Dove*, owned two (chauffer-driven) Navigators—in bright red."

UniWorld's strategy was to target the African American consumers across a wide communications platform that included television, public relations, events, and print media. One of the first initiatives of the Navigator launch was to partner with *Savoy* magazine, where UniWorld had been an advertiser. Together with *Savoy* and the American Black Film Festival, the agency created the "Black Oscars"—an invitation-only event attended by Hollywood tastemakers. It was where the Navigator was showcased.

Additionally, 85 percent of African American influentials say that once they find a brand they like, they stick with it. In fact, a Yankelovich study shows that 64 percent of African Americans will pay more for the best, compared to 51 percent of Caucasians, making them 25 percent more likely to buy premium or luxury brand names.[2] Consequently, their expectations of how they are treated both in media and personally are significantly more demanding. They continually evaluate both advertising and retail environments for their attention to and appreciation of AAA buying power—and what they discover profoundly impacts their loyalty to the brand. So, what are the two key takeaways here?

- AAAs are wealthier than the major market, and *in a stage of life that promotes even more consumer spending.*
- AAA are very brand loyal *when a brand has earned their trust.*

LANDSCAPE OF OPPORTUNITY

How do you, as a marketer, attract the attention of AAAs and earn their loyalty? When Andrea put asked Soledad O'Brien, producer and narrator of the CNN special Black in America, this question, she said, "One of the things we learned in discussions around CNN's Black in America is...affluent and influential African Americans are the target audience most major brands are desperately trying to reach. This elusive yet powerful segment not only has voting power, they have spending power. When this niche audience is satisfied, it shows in increased sales and loyalty."

How do you satisfy this niche? In July 2008, Diversity Affluence and *Uptown* magazine collaborated on a survey that revealed insights any marketer with the goal of reaching this audience can use:

- AAAs are very social. This group is very interested in the fine dining and social scenes, and is always looking for insider information on the newest and best in town. Nearly half (47.8 percent) go out for fine dining at least once a week. Over a fifth go clothes shopping at least once a week, and over a tenth travel on business at least once a week.
- AAAs are on the move. They travel frequently, both domestically and internationally, and often surround themselves with luxury items and experiences along the way. As can be expected, travel within the United States is most frequent, but international destinations like Canada, Europe, the Caribbean, and Mexico are also popular and common. Over 70 percent have a passport and have used it on international travel in the past year—with about a third traveling internationally at least three times a year and a tenth traveling internationally at least every other month. When asked how frequently they used their passport in the past twelve months, 22.3 percent said three to five times, and 41.1 percent said one or two times.
- Luxury shopping is a common activity when traveling. And even about a quarter fit in luxury shopping on business travel as well. When asked: How often do you shop for luxury items when you travel principally for pleasure, 46 percent said "frequently."

- The entrepreneurial drive is strong among AAAs. Given the necessity of networking to successful entrepreneurship, their strong interest in keeping current on the social scene may not be just for social reasons; the drive to keep up with the social scene may also be rooted in the drive to establish and maintain a successful business network. Nearly 30 percent of respondents make the majority of their living through entrepreneurial activities, and another 39 percent aspire to do so in the next five years. Another quarter currently have significant entrepreneurial activity.
- Vehicles are an important part of the current luxury experience, and they are looking to trade up even more in this category in the future.
- As a group, this market is quite independent given the high propensity to be single/divorced and not to have dependents.
- Fitness and wellness are an important lifestyle activity. While over 60 percent have gym or fitness center memberships, nearly 40 percent wish they were doing more to stay fit. About a third have home gyms, and almost 14 percent have a personal trainer, which is indicative of a rising trend in fitness routines across the board.
- AAAs like to be influencers in their social networks. But, further, they see themselves as resources for friends and colleagues on what's new. Research shows that, positive or negative, word of mouth spreads faster among AAAs than within the general market. African American influentials are well-networked, as they speak to an average of fifty-six people daily—a full 40 percent more people than non-affluent African American and 20 percent more than other U.S. influentials. But we can imagine that "shout outs" about a pleasing product or retail experience—or one that is displeasing—will also travel in the future by text and e-mail more often as well.

INSIGHTS FOR MARKETERS

Higher-end department stores are the retail venue of most frequent choice, but better specialty chains are also popular. Reinforcing that even affluent shoppers look for good bargains, however, outlets and "last chance" stores are also quite well-liked. Almost three quarters of AAAs say they shop most frequently at specialty department stores such as Neiman Marcus or Saks Fifth Avenue, and almost 52 percent seek out "last chance and discount" outlets such as Off Fifth or Neiman Marcus Last Call, or Gucci and Ferragamo Outlet Stores.

AAAs are savvy shoppers and research luxury items extensively before purchasing, so online information resources are a valuable tool in attracting these customers. There was also a strong local component to the information respondents accessed, or would like to access, to enhance their luxury purchasing experience. Local print media was used most notably for information on luxury apparel. Local or regional online media was preferred to identify information on special events/fundraisers and fine dining.

- Value is important to this market. Retail store atmospheres are most successful when they are warm and inviting. We found that although AAAs do not care to be confronted by a lot of sales/discount point-of-sale materials, most research participants were interested in getting some kind of deal or discount. But "discount" to this group still means luxury, just "affordable luxury." Everyone wants to stretch a dollar. A great buy to the AAAs is equivalent to purchasing an expensive item, but not paying full retail for it; for example, a pair of $800 Valentino leather pants for $400 at Neiman Marcus Last Call is what appeals to AAAs. They live by the adage, "It's not about how much you make but how much you keep." Lastly, they appreciate a personable salesperson to direct them to understated sale items rather than flashy POS.
- While some "mainstream" luxury magazines like *Town & Country* are beginning to cover Black-oriented social and media events—a trend that is both pleasant to observe and reflective of smart marketers responding to a shifting demographic—and while research shows that AAAs do consume general market media, they also consume Black media. In focus groups we conducted on media preferences, one interviewee stated: "I read the *New York Times*, *Town & Country*, *House Beautiful* and *Forbes*, but I also read *Essence*, *Ebony*, *Latina*, and *Uptown*."
- A common belief is that you can reach all diverse markets, in particular AAAs, through general media outlets. There is an assumption that once you reach a certain level of affluence or status the high end will to appeal to all, and that there is no discernable distinction in

the market. This is false and this point is so relevant we are going to say it again—this is false. It dismisses the nuance that is so important to marketers in winning the trust and loyalty of the AAA audience. And trust and loyalty go hand in hand with recognition and respect. By participating in media that is directly relevant to the AAA market you show that you acknowledge the market, and that you want it.

Our research also shows that this consumer is seeking more intimate, smaller scale social and business events that incorporate wine and wine education with a philanthropic overlay. A party with a purpose! What does this translate to for wine marketers? How about invitation-only wine dinners to promote your new vintage? Better yet, how about including a philanthropic overlay by choosing a charity that is meaningful to the AAA audience as the beneficiary of the event's philanthropic aspect? In 2009, we read an article in the *Wall Street Journal* about how "The French government is sponsoring cocktail parties in 19 countries across the world, including the U.S., Canada and Spain in an effort to boost sales of wine and cheese—two of its more lucrative exports."[3] All we could think was that it represented a perfect example of a missed opportunity: to target AAAs by partnering with a Black-oriented, member-based or philanthropic organization and niche media outlets to engage this consumer—*a consumer who already has an affinity for the products the French embassy desired to promote.*

FORGING THE BOND

The results of a focus group Diversity Affluence conducted in Miami help to underscore some of the most salient points we've made in this section. The group consisted of eight men and two women, between the ages 28–40, all of who qualified to be considered an AAA in terms of income. This group of AAAs:

- Prefers simple design and packaging when it comes to branding.
- Seeks out limited edition products because of the product's perceived exclusivity; too much advertising makes the product or

service undesirable as it implies mass production (the opposite of
exclusivity).

- Purchases one brand over another because it "do not want what
  everyone else has." Or "To set myself apart."
- Requires both quality and value in the goods and services it
  acquires—and luxury by definition means quality and therefore,
  value.
- Told us, when we asked the members of this group how luxury marketers could better reach them and their peers, that marketers should
  focus on exactly the sorts of ideas we have been promoting.
- Appreciates improved customer service, mainly at retail.
- Responds to brands that give back to the community by way of
  philanthropic support and marketing partnerships with member-
  based organizations.
- Believes that marketers need to gain a better understanding of the
  values of the Black community.
- Would like to see more marketing programs that incorporate a
  "generational" tone: that is, passing down values, assets and financial gains to your children and family.

The key learning from this focus group can apply, certainly, to all
luxury brands in any market. But the key to applying these points to a
brand strategy that targets the AAA market is in understanding your
brand's DNA and its relationship to the AAA customer. Sometimes
this requires creating relevancy, and other times a basic understanding of where and how your customer perceives your value.

MEASURING SUCCESS

Today, marketers and business development professionals are looking to emerging markets, such as Dubai, to grow their business. But
what about looking in your own backyard? What about making
your domestic marketing mix more robust, well rounded, inclusive,
and profitable?

It's important to make sure that any marketing outreach that you
do, no matter how small, is measurable. We recognize that measurement is not always attainable in every campaign, but in order to

begin any targeted market outreach you have to understand your expectations and then measure to that. What we believe will happen is that you will receive a pleasant surprise. What follows is a case study of how one luxury brand measured its success when it tapped into the AAA market.

An iconic French luxury brand, when it was doing quite well in the marketplace, wanted to expand its reach to new customers and be more inclusive. It is a good example of what we call a brand being proactive versus reactive. The company worked to develop a pilot program that would show measurable results with this market and would usher it into this new territory. It decided to form a marketing partnership with an established Black cultural arts organization that already enjoyed the enthusiastic support of African American philanthropists and business people. Marketing partnerships—mutually beneficial relationships between a brand and an organization in the non-profit sector—have been traditionally "one-off" affairs. A company may underwrite a dance troupe's tenth anniversary gala, gaining one night of gratitude from the troupe's regular fans, but going home as soon as the applause dies down and the champagne bottles are recycled. But what the French brand's executives learned is that establishing long-term relationships are key to fostering loyalty to the brand by people who are already loyal to the charity.

There are many considerations in making a choice of a long term marketing partner. In the case of a luxury brand the most vital is matching the integrity of the work produced by the non-profit with the integrity of the brand. For this French brand, it decided to focus on a city cultural institution, a dance company, which itself was a step out of the box in terms of marketing to AAAs. When marketers have thought associating with a charity to attract this segment, they have traditionally considered health organizations that focus on health concerns, Black rights or churches first. While there is certainly nothing wrong in affiliating with such groups, the affinities of

AAAs are broader than that, and appealing to these diverse interests can help your brand step out of the realm of the ordinary. As a first step—again, easing the brand into the new marketing strategy—it was arranged for the French brand to host a three-hour, invitation-only shopping event at its flagship store in New York City. The guest list was composed of supporters the dance company culled from their roster of "top" supporters, and the brand issued invitations to 650 of them.

These invitations were designed, produced, and mailed by the brand and, in keeping with its image, were top-notch. The paper was beautiful to touch as well as to look at, and each envelope was hand addressed. The actual invitations were so lush that just a day or two after going in the mail, invitees began to RSVP with requests that invitations be sent to their friends who had seen the invitations but hadn't received one themselves. This response gives us several insights into the community we're trying to reach. First, when it comes to branding, research shows that this consumer prefers simple, sophisticated design and packaging. These invitations were an extension of the brand's customary high-end style and therefore appealing on an aesthetic level.

Second (though not secondarily), the AAA consumer has a high propensity toward limited editions and one-of-a-kind products and experiences. In focus groups convened by Diversity Affluence, participants were clear: Responses included assertions of the fact that they "do not want what everyone else has" and that they were drawn to brands that "set me apart." This propensity stems directly from the youth experience of many AAAs in being a part of the urban culture movement. This was a group that as young people absolutely set the trends in fashion, music, and movies and influenced popular culture. As they have matured, their tastes have become more refined and their pocketbooks more powerful, but they have not lost either their ability to forecast what will be in fashion, or their desire to be first adaptors.

The beauty and exclusivity of the French brand's invitations were already creating a buzz, and making the invitation a coveted one. The shopping event was designed as a benefit to the dance company—ten percent of all sales would go directly to the cultural institution. That is, the more luxury goods to which the invitees treated themselves, the larger the donation the charity, to which they were already patrons, would receive.

Two hundred and fifty people attended the shopping event at the brand's flagship store. Within three hours more than $85,000 of merchandise was sold. A true and enduring partnership was established, with the brand's next gesture being to host a private dinner for the dance company's board, select other sponsors, and prospective supporters. Gift boxes greeted each guest, and an executive from the company welcomed everyone personally; the company has now expanded its involvement by creating a custom apparel item in celebration of the dance company's important anniversary where a percentage of the proceeds from the item goes back to the organization.

With the campaigns we will propose in the balance of this book, you will be able to take accountability to a whole new level.

We will also continue to prove the point that in many cases the AAA consumer is no different from the general market in that they have wants, needs, and aspirations. And marketing to them isn't difficult. In many instances similar approaches you've taken with other affluent audiences apply to AAAs. The primary differences are often found in the marketing mix, tactical elements, and frequency of your outreach. The question is, how do you accomplish the creation of brand affinity and loyalty with real budget, operational, perception, and human capital challenges? Well, the simple truth is, all marketing departments and directors have challenges when the desire or need to reach new markets become apparent. Where am I going to find the money in the budget? How will I convince upper management to take a chance on this, and do I want to take a risk with my job by advocating for the unknown? Do we have any-one internally to handle one more responsibility and, if not, can we afford an outside vendor or consultant? To add to this conundrum is the fact that the average lifespan of a CMO is about two years. So taking risks, while trying to navigate corporate bureaucracy and keep your sanity has its own set of challenges even if the audience and opportunity is viable.

With the vast number of media choices available to consumers today, all bombarding consumers with thousands of messages on a daily basis, breaking through the noise in order to make a brand impression or motivate the buyer to purchase what you are selling is no simple task.

In speaking with the president of that French luxury brand—who believed that the AAA audience was an important and via-ble one—we found that he was afraid to go back to his European bosses and ask to shift money from an already approved budget to take risks with new media and new markets. "It's just too risky right now," he told us. Another client—a non-luxury but well-known children's brand—sought us out and said, "I have no proof that reaching out to AAAs is right for my brand." But intuitively, he knew that there was business out there that he was missing; he wanted his fair share of it, and he wanted us to talk him through the risks and propose options that mitigated it insofar as possi-ble. Through marketing audits of still other clients we discovered under-tapped equities and assets they had already invested in but didn't realize applied to marketing to this specific audience. Via a marketing audit, one automaker, for example, already sponsored several programs with financial contributions and car displays at AAA events but they didn't incorporate a product specialist, data capturing or any prospect relationship management (PRM) sys-tem. While it is encouraging to see such forward thinking and movement, their ROI is lower than it could or should be without a follow up strategy. By creating a standard operating procedure for all programs that cultivate prospects, your rewards will be exponentially greater.

At the end of the day, this is the most important question: how do you mitigate the risk of outreach to what may be an entirely new audience while simultaneously increasing your ROI?

It's clear that affluent African Americans are low-hanging fruit, here in powerful numbers and with dramatically rising influence. The change within the Black community—its demographics and psychographics—is seismic. So how to capture this opportunity? Progressive marketers, savvy entrepreneurs, brands, and non-profits can take three simple—yet powerful—steps to create loyalty among African American Royaltons: (1) acknowledge them, (2) understand them, and (3) cater to them.

CHAPTER 4

INSIGHTS, TRENDS, AND 360-DEGREE MARKETING

THE TWOFOLD CHALLENGE

AAAs are, in general, an information-driven group. But like all consumers, they get their information from a variety of sources—and they access this information in ways and at speeds that were unheard of just a few years ago. Whereas once a listener might have tuned into a favorite local radio station to catch its hourly news broadcast, today's formats and technologies—including talk and satellite radio—provide a wider range of choices. A decade ago most of us read our hometown daily newspapers after we'd retrieved them from the front stoop where the paperboy had tossed them; today we can—and do—access nearly every newspaper in the world from our laptops and Blackberries.

Whether our primary news source is the radio, the Internet, or one or several newspapers depends on how easily we can access the source as well as our personal preferences. There is no one form of information that any of us, including AAAs, turn to en masse, just as there is no longer one television show, like CBS Evening News with Walter Cronkite in the old days, that a marketer can depend on to reach any sort of a general market.

The simple fact is that there are more ways to reach out to more people than at any other time in history, but the technologies that allow for this reach are so new that most of us haven't figured out yet how to use them to our advantage. The challenge this presents to marketers is twofold. First, we have to figure out how to creatively separate our message from the herd's—how to make the value of our product or service stand out in the confusing cacophony of media voices that vie for customer attention. Then we have to dive back into the media morass and decide which venues most effectively communicate our message.

TRENDSPOTTING

The phenomenon of rapidly advancing technology is combined with a shift in consumer spending habits and emerging affluent markets. This shift may feel quite subtle—indeed, to some it may feel like the temporary result of an economic downturn that will soon be over, allowing consumers to return as quickly as possible to their old rapacious ways. Trend forecasters like Andrea are already warning us that this is not so; consumers are certainly still buying, but they are now buying in ways that are quite different from the recent past. In this climate, the usefulness and urgency of knowing who your potential customers are and how they define value cannot be overstated. Below are the top AAA consumer trends that we are predicting:

1. **Affinity for Autos Especially for AAA Women.** In the first half of 2009, according to R.L. Polk, 12,606 luxury autos were purchased by AAA women. One way that corporate can take advantage of this emerging segment is to start women's driving clubs, creating a pilot program in a region with a high concentration of AAA females and holding monthly "ride and drive" outings. These events can be paired with an online component, perhaps a social network for a particular women's driving club, or a charity. But don't assume that these go-getters want "girly" outings. They just might surprise you and like the same driving experiences that men would—fast and furious! Don't assume you know what your audience wants. Ask them.

2. **The Experiential Vacation Less Traveled.** It's not just fluctuating fuel prices that are keeping people home. Home and hearth are taking on new meaning as people begin to place more value on the experience that their money can buy as opposed to the material goods it can acquire for them. This trend is good news for museums, historical sites, and other places of cultural interest. But what it means for marketers of more exotic resort and vacation destinations is that they'll have to work a little harder (and reach out a little farther and more intelligently) to demonstrate the experiential value of what they offer. However, the good news is that this consumer is interested in using vacation destinations to connect with friends and family. Annual events, such as fraternity or family reunions, and events for affinity groups, such as conferences for member-based organizations, take on new importance as people seek connectivity. Additionally, our research indicates that women organize "women's vacations," and men do the same. They are looking for unique experiences that provide knowledge and value to their lives and friendships and help them relax while being treated extremely well. We learned that spa destinations, golf and ski resorts, or trips to places with a connection to their culture, such as African safaris, are on the AAA radar. A key takeaway when marketing and promoting to this audience is to a limited time discount. Remember, AAAs look for value and respond favorably to exclusive discounts.

3. **The Art of Cooking.** Here we're not referring to just the average consumer who is trending away from eating out and toward the cooking and baking of traditional "comfort foods" at home, though that has certainly been demonstrated in any number of marketing reports since 2008. The upscale diner is also trending homeward, toward dinner parties for select friends, family, and business associates. The luxury marketer needs to rethink how his or her product can be positioned to enhance the home dinner or cocktail party. The new luxury consumer is going to need and want everything

from unique recipe ingredients and fine wines to beautiful flowers, crystal, china, table linens, and maybe even a new table, so they can pull off home entertaining with signature style. Former model and restaurateur B. Smith now has a line of home goods. Luxury marketers would be smart to find similar collaborations to create new or expand *existing* product lines.

4. **Organic.** In chapter 5 we will go into more detail about the forecast for the way the new green consciousness is going to impact buying habits within the AAA segment, particularly in relation to vehicles. But for now, just note that while organic products usually wear steeper price tags than their nonorganic counterparts, wealthy consumers retain their ability to afford these higher priced, premium products. They are also among the consumers who make taking care of their health their top priority, which means this elite group has been increasingly embracing the health benefits of organic foods. The concern for personal health and well-being doesn't stop with organic foods; it extends to other green products as well. From beds to bed linens and cosmetics to cars, the informed luxury consumer is going to start looking more closely at the environmental quality of all of the products he or she buys—and at the greenness of the companies who make them.

5. **Content is King.** AAAs watch TV, listen to the radio, read magazines and newspapers, and spend time online. And they want content that resonates with them. One focus group participant indicated that while reading *Travel and Leisure* magazine, she came upon an interview with a Black celebrity who had visited a certain island. She was so impressed that she quickly made plans to visit that island herself. This is a perfect example showing how general media was more inclusive with their content and it delivered results. Aligning print, broadcast, and online advertising with venues that provide the content AAAs are attracted to (magazine articles, documentaries, Web sites) is the way to go.

6. **Style and Status.** Affluent Black males have long been underrated, but style and status are important to this segment of the market because they are highly aspirational, and they view themselves as influencers and role models. This consumer group is an early adaptor of premium electronics and new technology, as well as autos and premium clothing from designers such as Zegna, and even custom clothing. Because of his desire to be distinctive and to wear clothing that fits well, the AAA man seeks out custom clothiers—knowing that custom-made clothes fit better and that they would ultimately pay the same price for designer clothing anyway. This segment-within-a-segment thrives on anything that is limited edition or one of a kind, particularly invitation-only events that gain them access to exclusive product launches and experiences.

7. **Parties with a Purpose are on the Rise.** AAAs across the country have expressed a major common interest in branded events that incorporate four key elements: business networking with like-minded peers; opportunities for personal networking; relevant charity to be the beneficiary; and, believe it or not, dating opportunities, if they are available. Because AAAs are time constrained, they value getting all they personally and professionally desire in one night. For marketers, this means ensuring that there is a good balance of men and women at an event, and ideally, that many of them are single. The evening's program should include content that's meaningful to the AAA, such as speakers who inspire them and encourage an evening of open dialogue. In the case of events, bigger is not better; smaller and more intimate events with one hundred attendees or less are more successful. Markets that would benefit nicely from hosting or being involved in such events include fine wine, fine dining, art collecting, affordable luxury travel, and spa resorts. Cheap wine, hard liquor, and light food need not apply.

8. **Relax, Release, Relate.** This one is simple. AAAs work hard to obtain and maintain what they have but, like most other busy

people, need a strong call to action to force them to act on their desire to have a healthy dose of work-life balance. Fitness and wellness is an opportunity to reach AAAs. Remember, while over 60 percent have gym or fitness center memberships, and about a third have home gyms, nearly 40 percent wish they were doing more to stay fit. According to focus group research, for both women and men, spa treatments in the local market were like a brief vacation—a quick "it's all about me" experience. Branded wellness resorts or "events," for example, could be an annual "Elizabeth Arden Wellness Retreat" or the "Calloway Getaway"—these would be the perfect prescription here.

9. **Low-Hanging Fruit.** We've all heard the rap lyrics talking about patron sipping "in da club," but those stereotypes are simply not accurate when considered outside the very narrow arena of rap videos. Upscale Black consumers have much more sophisticated palates and are looking for the best in fine wines. AAAs want to see and associate themselves with a broader vision of the way African Americans can be successful that goes beyond music and athletics. So malt liquor ads, cognac ads, or celebrity endorsements of tequila or vodka aren't what appeal to this consumer. What does? High-end wine advertising, and content related to collecting, consuming, and entertaining is welcomed and encouraged. Wine-store events or invitation-only wine dinners hosted by the vintner will also be popular.

The booming trend is wine consumption among wealthy African Americans, who annually spend $4.4 million on in-store wine purchases and are more likely than the general population to spend $20 or more per bottle. According to our 2008 research, among AAAs, 30.8 percent reported that they consume imported wine at least once a week, and 31.3 percent consume domestic wine at least once a week. This consumer group also expressed a deep interest in attending invitation-only wine dinners where they can learn about wine and wine collecting while connecting with like-minded peers.

Note to wine marketers: there's a huge market here, but when you decide to host your invitation-only wine tasting, be sure to play some soft jazz, but hold off on the rap for the evening.

10. **Innovators and Risk Takers.** As we've already mentioned, the numbers show that AAAs are starting or expanding businesses in growing numbers. Marketers have a sizeable opportunity to reach those living the American Dream by tapping into business affinity groups, publications, and annual events that appeal to these self-made prospects. Content in on- and offline publications focused on Black entrepreneurs creating wealth and success might just do the trick. And these visionaries are likely the same folks you want to invite to sit on your corporate or philanthropic board. They can offer advice, guidance, perspective, resources, money, and their potential contributions should not be overlooked.

So, how do we use these trends? How do we adapt what we know about the interests and affinities of this segment and put it to use in our targeted marketing campaign? That's our final challenge, of course: adapting our marketing skills to focus on niche segments, because the traditional ones just aren't what they used to be. They're stagnant, or they've reached a saturation point. We can no longer rely on conventional customer bases alone—not if we want our business to grow. Understanding shifting demographics and tapping into expanding ethnic markets such as affluent African Americans is crucial to not just maintaining but expanding your brand's bottom line. But how do you get your product or your service inside this tightly networked niche? What are the specific challenges to tapping into the AAA segment? How do you find their sweet spot?

Decades of experience and countless marketing campaigns with major purveyors have proven to us that a 360-degree approach is what it takes to engage today's consumer, and in this respect the AAA audience is no different. Brett Wright, co-founder of *Uptown* magazine concurs, "Since *Uptown*'s inception, we've experienced

a growing demand from luxury advertisers to add more marketing platforms to our media offerings, and we've done just that with
Uptown 360. It's great to be working with some of the luxury industry's most savvy marketers who view us as a true marketing partner with access to an audience of qualified prospects. An integrated
marketing communications media buy allows a brand to tap into all
of the consumer touch points, and that's what *Uptown* delivers."

A 360-degree approach is defined as an integrated multime-
dia strategy. But, according to a recent report by the American
Association of Advertising Agencies, although 91 percent of senior
marketing executives believe that this kind of integrated approach
is critical to their success, just 21 percent are comfortable that their
organization actually does a credible job delivering it.[1]

We believe part of the reason for a lack of good results when
deploying a 360-degree approach is that many people see it as a
complex undertaking and manage it inefficiently. Some even go so
far as to argue against the approach, insisting that marketers should
choose to focus on just a few of the many media channels available
to connect with consumers—the "do one thing and do it exceptionally well" method. You'll get no argument from us that each portion
of a 360-degree plan must be executed exceptionally well. But that
is not enough. To really succeed in this strategy you must reach out
to all the media outlets your target market consumes. That doesn't
mean do it alone; partnerships can help provide not only value (that
is, lesser initial expenditure), but you can also leverage their own
brand equity and assets to complete your integrated program.[2]

We can't help but believe that this tendency to shy away from
360-degree marketing is a response to its seeming complexity. In
order to reach their traditional customers, marketers fall back on
what they know, blinding themselves to the new markets. And as
a result they are leaving whole sections of a market out—the AAA
market that, as we have shown, has the financial ability to be a key
driver of their business growth.

Yes, coordinating your company's magazine ads with its special events and its presence on social networking sites with press releases to appropriate Black media outlets is more involved than in the old days, when you made your annual buy at a newspaper and augmented the campaign with a few regular thirty-second radio spots. But those who are still yearning for seemingly simpler times need to acknowledge that those days are gone, or their brands are going to vanish from the consciousness of the 360-degree consumer.

The 360-degree AAA consumer? Stripped down to its essence, the unavoidable fact is that marketing is no longer company-based, but consumer-based.

For no other audiences is this truer than in the affluent ethnic market segments. Part of the reason for the "show me" way of thinking—and we will venture to say that it's a *big* part—is that this audience has been and continues to be neglected as a viable and influential target by luxury marketers. This neglect manifests in many different ways—and by the time you're done reading this book you'll know them all—but let's start with the fundamentals: the AAA consumer has basically gone generally unacknowledged by luxury brands. Companies are happy enough to allow an AAA to spend his or her money on their products or services, of course. But proactively reaching out to and welcoming the potential consumer as a valued customer, literally and metaphorically throwing open the doors with a full understanding that she and her AAA fellows want to and will and *do* consume on a regular basis the fine things such as any one company has to offer—well, that just hasn't been a habit of luxury brands.

But part of the reason is also that the AAA segment is merely accustomed to being 360-degree customers. Let's not forget what we mentioned earlier: this segment may no longer be motivated by urban and hip-hop references but, back in 1991 or so, many were indeed part of the urban youth culture—and the urban youth culture was marketed in a full, fat 360 degrees. Their own radio and television stations, their own magazines, their own movies and fashion, their own events where they could gather with like-minded peers—the AAA segment has gotten used to being marketed to in a lifestyle manner. And now that they're grown up, they are still waiting for marketers to keep up with them.

At the end of the day, however, AAAs put their designer shoes on one foot at a time, the same way everybody does. They have the same desire to provide for their families—a beautiful home, or two; a first-rate education for their children; regular getaways to desirable locations. And they want to be recognized as living the American Dream in the same proportion as they have worked to attain it. *The brands that best recognize their achievements will get the bulk of their business.* That's a key statement, and it revolves around basic human psychology: it's a natural instinct to show gratitude to people who go out of their way to speak to you and include you in the conversation.

So how does a brand go out of its way to speak to the AAA audience? What does it have to *say*—beyond, perhaps, what the brand is already trying to communicate by including African Americans in the creative or putting a Black mannequin in the store window? What can a company—perhaps one that already ascribes to a 360-degree marketing viewpoint—do to enhance their status within the AAA group? Maybe your brand or business already has investments and equities among this consumer and you don't even realize it.

Here's a paradox for you. The answer to those questions is this: a brand can do many of the same things it has always done to reach what was once its conventional market segment. It is often merely the small nuances added to an existing plan of action that can make the difference.

HSBC Premier is a global banking service designed for affluent and internationally mobile consumers. It's a company that is accustomed to serving diverse market segments at the global level and to providing its well-traveled, well-educated, socially aware and globally sophisticated audience with a high level of personalized service.

So, when HSBC wanted to make the AAA segment feel more at home with the company, it assessed its consumer outreach and tailored what it was doing to fit this group.

Given its heritage of serving diverse audiences, it's not a surprise that HSBC came to the table to develop its AAA target plan with its own name for the group—"AH," or the African Heritage segment. "We find that term more inclusive than African American," Armand explained, "because it includes those from the Caribbean, South America, etc. We see it as a broader approach. When we looked at that landscape, the AH market, we saw it was treated in a homogeneous manner, particularly in the financial sector. With HSBC Premier, we wanted to make sure that AH was included. We recognized that there's an affluent population in that community and there were opportunities to raise awareness in that segment."

What HSBC found in its own independent research was that taking advantage of those opportunities would require tailoring its marketing outreach program to highlight the features that the AAA group had indicated were important to it. And they noticed something important—that arts and culture was a very effective way to reach African American Royaltons. One example of how the insight translated into action is that HSBC began partnering with Evidence, A Dance Company, participating in the troupe's winter gala. It was a "bespoke" approach, according to Armand, who is very selective about lending HSBC's name to such partnership programs. Does the program resonate with HSBC's target audience? Can they develop a long-term relationship that will continue to build momentum and traction and goodwill, rather than participate in one stand-alone event that will come and go and likely be forgotten? Armand confirms that in this case, the relationship has been a stunning success: "[Evidence] incorporates music from the African, Latin, Caribbean and urban communities.... People who attend Evidence's performances...are more attuned to who we are. They like culture and they travel."

In addition to supporting art that resonates culturally with their target audience, Armand also knows the importance of backing up a signature event with participation in the media the segment embraces. As we've already talked about, there is a perception that once diverse consumers get to a certain level of wealth, they assimilate, and that perception, to a degree, is true. "But while someone might read the *Wall Street Journal*," Armand explains, "I feel that people who have specific interest, or who belong to specific communities, have a higher engagement level with other media outlets that speak to those interests more directly. They're more truly enthusiastic about them. That's where Premier needs to be present, beyond the mass market. The hard part is identifying what that media is. The point is that there is a cultural affiliation that we recognize: if I'm interested in responding to your needs, oftentimes I'm better at being able to do that with authentic, niche media."

Even more plainly: if you're a manufacturer of golf attire, you could advertise in *Vogue* or *GQ*, but one of the magazines that targets golfers is a much better bet for finding people who have a natural affinity for your wares. So, if you want to sell to affluent African Americans, you want to be in the magazines affluent African Americans read. Unfortunately, there are not a lot of these print outlets; *Black Enterprise*, *Essence*, *NV*, and *Uptown* are your primary resources. When Armand wanted to add another piece of the 360-degree pie to HSBC's AAA outreach, the company partnered with a Black publication for yet another signature event, a culinary evening at the HSBC's headquarters in Manhattan.

At this point in our description of HSBC's marketing plan, you might be beginning to wonder at the cost of all of these print ads and specialized, signature events, so we want to address another aspect of 360-degree marketing that can make some people resist the approach—and that is *expense*.

When we talk about marketing budgets and the costs involved in producing and executing an effective 360-degree marketing plan, we do so without consideration for the general condition of the broader economy. In any economy—boom or bust—the difference between a successful marketing plan and an unsuccessful one is efficiency. In a bust economy one has to make every marketing dollar meaningful, certainly—but is there any reason to waste those dollars in boom times? What makes the AAA market right for this economic time is that grassroots is still an important and meaningful tactic to reach the audience effectively. And when you compare this to a brand's traditional means of reaching customers (such as the *Wall Street Journal*, *Financial Times*, and *Vogue*) we feel that reaching out to the AAA audience can reap significant rewards, especially in relation to the costs of acquiring these customers. "Marketing Directors are increasingly charged not only with justifying [expenses] but actually taking ownership of the P&L."[3] How can you measure that return?

In presenting an opportunity for a diversity initiative, we often encounter clients resistant to moving money around within their existing marketing budgets to try something new. They are accustomed to, for instance, advertising in the aforementioned *Wall Street Journal*—newspaper ads being a particularly old-fashioned approach that will not make in-roads to reaching a new market. But when we ask them what return they see from the ads they place in that publication, they often don't have an answer. They run in the *Journal* because they've always run in the *Journal*, and everyone else runs in the *Journal*, so of course they believe they have to continue to run in the *Journal*, too.

"If we can take the budget you have for just one or two of those *Journal* ads with an open rate of 60K for a half page," we tell them, "and repurpose that toward a pilot program or two and show you a positive ROI, would you be interested in testing a program to target this audience?" At that point, many clients will agree. So the first step is to conduct a marketing audit to determine which assets can best be leveraged for a pilot program. Perhaps we will find that the company in question has a long-standing tradition of hosting

invitation-only events to benefit a certain philanthropic organization. In this case, we might suggest that the company simply host
another event—but broaden its guest list by partnering with a
Black-oriented, member-based organization and choose a charity
that has particular resonance with the Black audience. Say, an intimate dinner party for sixty guests culled from the top patrons of
a Black cultural arts organization. We would advise the brand to
stage this sort of event in several cities—New York, Washington,
D.C., Atlanta, and Houston—over a one-month period. At the end
of the pilot program, the ROI would be examined based on previously agreed upon measurements—for example, by how many the
brand's prospect list was increased or, if the event included an in-
store sale, the number of units sold. Based on our extensive experience, the brand is likely to find that for the $120,000 that it would
have cost them to run two half-page advertisements in the *Wall
Street Journal*—a marketing method effectiveness of which is essentially untrackable—they have bolstered their prospect list by over
300 flesh-and-blood new customers and/or sold an amount of hard
goods in the course of their four or five intimate events that well
exceeded the cost of the events themselves.

We are standing at a crossroads in the marketing game. Most
marketers have, until recently, never really been asked to measure
the results of their advertising programs, especially to the degree
that we have the ability to do today. But more than ever before marketers are being required to take responsibility and show results for
the dollars they spend. Accountability has come out of the closet,
and regardless of the fluctuations in the broader economy, it's not
going to go back in. There are too many ways these days to measure almost anything, and little reason not to. And the bottom
line is, if you're not moving the needle on your company's bottom
line, you're not doing your job. There are so many opportunities to
reach new markets here in the United States. Marketing innovation
needs to take centerstage, and we aren't referring to edgy-creative

but to affordable, measurable, profitable, and meaningful consumer outreach.

Almost invariably, when we present clients with the opportunity to measure ROI, they get excited. So now let us tell you how innovative outreach manifested in a positive ROI for HSBC.

In partnership with *Uptown* magazine and its thriving online community, the culinary event HSBC sponsored featured a tasting menu by a top AAA chef, Marcus Samuelsson.

Born Kassahun Tsegie in Ethiopia in 1970, Samuelsson and his sister were adopted by a Swedish couple when their birth mother died during a tuberculosis epidemic. He became interested in cooking because of his maternal grandmother and studied the art at the Culinary Institute in Gothenburg. In 1991, he came to the United States as an apprentice at Restaurant Aquavit. When he was just twenty-four years old, Samuelsson became executive chef of that restaurant, and it was in short order that he became the youngest chef ever to receive a three-star restaurant review from the *New York Times*. In 2003 he was named "Best Chef: New York City" by the James Beard Foundation. One night, in early 2007, Samuelsson also cooked for about 125 high-net-worth attendees at the HBSC event.

The evening was billed as a "Taste of Africa." The food had what Samuelsson told guests was a "Senegalese flair," incorporating native ingredients like peanuts, cous cous, bananas, sweet potatoes, ginger and *Buy*—the fruit of the native baobab tree—with the French influence found in many Senegalese dishes. And, in addition to an extravagant menu prepared by a renowned chef, the evening included the giveaway of a trip to Africa.

This successful evening was not out of the ordinary for any event hosted by a luxury purveyor for its prized customers. The only difference was that the majority of these customers were Black. The nuances were very small—the choice of cuisine, the giveaway trip destination—but they were appreciated by the audience HSBC had invited.

How did the audience show its appreciation?

A year and a half after the event, HBSC reported that, to date, they had credited 2.5 to 3 million in deposits to that one evening when they entertained 125 people with Marcus Sameulsson's gorgeous food.

And that's how to get a measurable outcome from a diversity initiative.

In this chapter, our purpose has been to underpin the importance of a 360-degree approach to reaching your intended target consumer. To show you that for the cost of a couple of ads in the *Wall Street Journal*, it is possible to add another whole dimension to your media platform—in this case, a culinary event for a few select guests—and show a grandly measurable return. We also wanted to give you a taste of the sort of nuances, or tweaking, that can make or break an outreach program geared for the AAA segment.

But we've covered in this chapter only two of the components of the 360-degree pie—event marketing and media. In fact, there are eight essential components of a true 360-degree program, and marketers should think toward creating a standard operating procedure when evaluating an opportunity and measuring and reporting on those that are chosen.

1. **Event Marketing.**   Luxury marketers have long courted their customers with special, signature events. The nuances involved in tailoring such usual occasions to a primarily AAA audience are quite subtle, and in no way add to the cost of a standard special event. For example, at the launch of a decidedly upscale perfume, the evening's entertainment was a performance by the Grammy-nominated R&B singer Emily King. The comments among the attendees were quietly approving, "[The company] knew enough to get her to perform. They get us. This isn't a cookie cutter event, it was planned with us in mind."

2. **Print Media.** We think it is impossible to overemphasize this point: put your ads where your audience is. The only publications expressly for the AAA audience are *Black Enterprise, NV, Essence,* and *Uptown* magazines. Remember that more is not more. While there are not a lot of luxury media outlets, the ones that exist are powerful and laser-focused. More importantly, partner with them and take advantage of all available opportunities to tap into their loyal subscribers and readers. Don't just "advertise."

3. **Broadcast Media.** Radio and television remain good ways to reach the mass AAA and general market. However, targeted buys on stations that program for AAA audience, such as Centric, TV One, and CNN's Black in America are still a valuable way to get the attention of this prosperous group.

4. **Philanthropy.** As we touched on earlier, giving back is important to wealthy African Americans. Companies that do the same are respected and patronized. Companies such as State Farm and McDonald's may do a good job of this, but luxury brands have the same opportunity. The majority of AAAs prefer to do their giving at the local level, and do it in their lifetimes rather than leaving bequests in their estates. Partnering with the right cultural, educational, or charitable organization can be a key in unlocking the goodwill of AAAs.

5. **A Seat at the Table.** This is an often overlooked component of a good overall marketing plan, so let us state the obvious: a marketer can break records getting consumers in the door, but if the sales and staff don't adequately reflect the audience, or aren't adequately trained to take care of the customer, then all that hard work—and all of the money placed against the segment—won't mean very much. All the planning and expenditures executed at the corporate level can come undone at the retail level if they are not managed correctly. Getting clients is one thing. Keeping them is another. Your sales and service staff are what keep your

customers coming back. Also, a seasoned consultant, advisor, or employee who is knowledgeable in this market should have a seat at the table when strategizing marketing efforts in order to guide you toward relevant touch points while avoiding mistakes and misperceptions. A seat at the table can also mean including more diverse executives on your advisory board. Or better yet, form a diversity advisory board to ensure they can steer your brand or business in the right direction.

6. **The Internet.** Research shows that the Internet is one of the most important business, shopping, research, and communications tools among college-educated Blacks, and its influence within this segment is only growing. Some popular blog sites include www. blackgivesback.com and www.theblacksocialite.com. Online-only publications include www.theroot.com and www.targetmarket-news.com. From online ads placed with media partners to using their subscriber lists to distribute promotional and invitation-only event E-mails, leveraging the Internet is a powerful tool today. Even making your Web site more search-engine friendly and incorporating content inclusive of Black models, your diversity initiative or diversity advisory board and the organizations you support are good starts. Repurpose print publication articles for the Internet and search optimize them. Reach out to bloggers to help spread the word about a product launch, or report on an event or donations made. Tapping the talent pool and becoming a sponsor of an AAA event, such as the Blogging While Brown Conference, would be a good start.

7. **Social Networking.** According to Lorenzo Benazzo, CEO and founder of Hyphen, a company that builds brand-focused social networking Web sites, "Social networking as a whole has been changing the way people interact online. But where it can excel is with culture or lifestyle-focused initiatives. That is particularly true if a company, under the umbrella of a recognizable brand, decides

to cater content, local initiatives, and services designed specifically for a culturally homogeneous audience.

"A brand-powered online social network can then provide a continuous tie among people who just met at a party. Conversations can start at a gathering and continue online. New relationships that were built locally can be nurtured via the Internet and grow with—at the center of it all and as the main tie—the brand that makes it all possible.

"When used properly, and within a more intimate setting than most major social networks, locally-focused brand-powered social networks can, and are becoming, the new way to promote a company and its offering. In addition, when it comes to luxury and high-end retail marketing, they can also turn out to become the ultimate marketing solution to bring people to a store and directly impact sales."

Though statistics for the impact of marketing online and through social networking sites are still hard to come by, we agree that these venues will grow only more critical to any modern marketing effort. This is just as true for luxury marketers. Some luxury brands are already ahead of the social networking curve. Burberry, for example, hosts the site www.artofthetrench.com, which it launched in an effort to "forge closer relationships with its customers and attract new shoppers."[4] In the article "Burberry to launch social networking site" in *Retail Week*, September 2009, author Jennifer Creevy went on to underscore the importance of social networking to luxury marketing, quoting chief executive Angela Ahrendts telling the *Financial Times*: "These might not even be customers yet. Or they may be a customer for a bottle of fragrance or for eyewear. But these are customers who need the brand experience, who need to feel the brand. That word-of-mouth spreads through their social networks and continues to be a positive conversation [about Burberry]...that is so powerful."

"Our industry is going through a cultural shift from ad time to real-time," Rich Gagnon, Draftcb New York's chief media officer said in a press release for the launch of the company's new group, Real-Time Marketing. The group combines its existing capabilities in digital marketing, media management, and customer relationship management. "At the heart of the 'Real-Time' approach is understanding how technology is affecting the relationships people now have with brands and how data is informing those relationships. Too often relationship management is viewed as a 'back-end' program versus an upfront strategy," added Mike Brzozowski, the executive director of CRM Consulting.[5]

But managing relationships through social networking media is becoming imperative, and luxury purveyors are beginning to understand this—some in a very public way. "Perhaps the rising power of social media to connect with luxury consumers was best expressed by Dolce & Gabbana in who was invited to sit in the first row at the designers' runway show in Milan last week. The heads of luxury retailing giants Neiman Marcus and Saks Fifth Avenue were ousted from their front-row seats by Internet bloggers Scott Schuman from Sartorialist, Garance Dore, and Bryan Boy, according to *Woman's Wear Daily*. Clearly, Dolce & Gabbana believe that the bloggers have more influence on their target customers than the retailers do. In other words, the designers are betting that it is better for the brand to have an outfit featured in a blog post with the potential to reach millions of eyeballs worldwide than in a Saks 5th Avenue window that will be seen by mere thousands."[6]

8. **Press Releases.** It seems like a no-brainer that a marketer would send out press releases to promote events and build a buzz around the products and services he represents. But did you know that there are dedicated wire services that will send your press release to over 40,000 journalists and bloggers within the Black news industry? In chapter 5, we'll tell you how to make sure your press releases are going to the right places.

Now, all of this said, every 360-degree targeted marketing initiative does not necessarily have to include all eight components. In the following chapter, we'll use case studies of successful plans to show you how to use each component in the combinations needed to make the most of your diversity marketing budget and grow your customer base. What's most important is reach, relevancy, frequency, and follow up. For some brands, AAAs will be a completely new and viable audience. For others, it represents an existing and perhaps undetected customer base that, with a correctly focused approach, can expand and increase profit margins dramatically—and even reinvigorate existing brands.

In chapter 5, we'll show you how to attain this focus and capture this market—not to mention how to create happy customers.

We leave you with this thought: it's all about *showing* your 360-degree customer that your product or service deserves to be part of his or her lifestyle. The rewards of doing so can be as rich as HSBC's dramatically increased deposits. And, if you choose not to do so, your competitors will. The question you should ask yourself is "is my brand or business a leader or follower?" Followers imitate but never duplicate, and playing catch-up to your competition is twice as costly.

CHAPTER 5

PUTTING INSIGHT INTO ACTION

SUCCESSFULLY MERGING THE EMERGING MARKET REALITIES

One financial firm, one evening, one world-class chef, one hundred and twenty-five satisfied diners: 2.5 to 3 million dollars in new deposits.

This is what happened when HSBC sidestepped Madison Avenue's clichés and reached out to the African American segment in a new, more meaningful, and measurable way. The needle moved on HSBC Premier's bottom line because, Nancy Armand told us, it "looked at that landscape, the African Heritage market, [and] saw it was treated in a homogeneous manner" and decided to take a "risk" on doing something a little different; something that reached out to this segment in a more specific and personalized way.

But how much real risk was there in implementing that program? You will find that the programs recommended in this book require more creative thought than time or even money to succeed. Think again of our challenge to repurpose the equivalent amount of money you might normally put toward just two advertisements in the *Wall Street Journal*. If you were really to put that money into newspaper ads that ran, for example, on any given Thursday and Friday, you could reasonably expect that come Saturday morning, those newspapers—and your advertisements—will be recycled.

If, however, you put that money into presenting just one high-caliber event, such as HSBC's evening with Marcus Samuelsson, you could be interacting directly with 125 happy and grateful potential customers on Saturday night—customers who provided their personal contact information to follow up with them. Come Monday, you would have the potential to add almost 3 million dollars of revenue to your business.

What we want you to do right now is to imagine stepping out of your marketing comfort zone the way Armand did. Envision creating new income streams for your business of the same scale Armand managed for HSBC. Picture that level of success happening over and over again for each brand, business, product, or client with whom you share the insights and knowledge you're finding in this book. By crafting cutting-edge, targeted campaigns, you will attract a whole new affluent market segment and make them into brand-loyal consumers.

This chapter is about the specific strategies you can use to find that same kind of success.

THE ONLY THING YOU CAN COUNT ON IS CHANGE

It takes a lot of courage to release the familiar and seemingly secure, to embrace the new. But there is no real security in what is no longer meaningful. There is more security in the adventurous and exciting, for in movement there is life, and in change there is power.

So says Alan Cohen, author of over twenty-two bestselling books, although probably best known for his contributions to the *Chicken Soup for the Soul* series.

What is it that's changing in the luxury marketing world? Where are tomorrow's leaders in the luxury goods and services industries going to find their power today? Many luxury brands are already blazing trails by finding new distribution points and creating more affordable entry-level products through marketing innovation. Let's review some of the changes that need to be embraced to fully shake off what is no longer meaningful in today's industry and move into the new world taking shape around us.

The first change we have to contend with is that the luxury consumer just isn't what he or she used to be. Pamela N. Danziger, president of Unity Marketing and a nationally recognized expert in consumer insights, especially in relation to luxury goods and experiences, had this to say about the marketplace in July, 2009:

> "... Unity's research shows that affluent consumers' basic attitudes and motivations that underlie their patterns of consumption are changing, and these changes are likely to outlast the economic downturn. The new survey points to opportunities for luxury marketers that align themselves with the new values-based mindset of luxury consumers."[1]

To be blunt, the luxury consumer is no longer buying just for the sake of acquiring or for the status conferred by owning luxury goods. People who are buying luxury are doing so because it enhances the quality of their lives. And AAAs are no different. That vacation in the south of France with a group of friends revolves, not just around eating fine food and drinking fine wine in an exotic location, but around spending a day with a winemaker in Côtes du Rhône—and maybe even picking Viognier grapes while they learn about winemaking. That new car isn't just another mode of transportation—it's a luxury hybrid model, and it taps into consumers' growing desire to be eco-conscious and to be seen as eco-hip. Like HSBC, that investment firm, isn't just about managing money; it also cares about the cultural institutions that add quality to the life of my community, and it finds inspired ways to support it. Remember, lifestyle marketing is one key to capturing this consumer.

The luxury consumer is, in short, growing even more discriminating. It's no longer good enough to simply make a high-quality product. Quality is expected these days. For insight into the thinking of this new luxury consumer, here are a few of the statements made by the participants in one of the focus groups Diversity Affluence conducted in Atlanta, Georgia. This group consisted of seven men and six women, ages 40 to 65, with an average annual income of $125,000.

- Service is a distinguishing factor as to how this group defines luxury: "The brand gets you there. The service keeps you there."
- They place a premium on all forms of top notch and consistently good customer service.
- "Luxury" means reliability, quality, value, and exceeding expectations.
- They prefer direct mail invites over electronic (e-mail) contact.
- Retail, retail, retail! What happens at the retail level is a huge part of the total luxury experience. This group feels that luxury is a personal experience that is made or broken at retail. You can win this group over with personal attention and by building a personal relationship at the retail level. Think thank you notes!
- The most effective way to gain this audience's attention is by delivering the total brand experience: advertising, event marketing, and philanthropic support. In other words, a 360-degree marketing approach.

It is clear you have to consider the total experience the consumer will have when he or she buys your product, and how that total experience will contribute to this consumer's quality of life.

The second change you will have to recognize is that your consumer demographic is shifting.

"The United States is undergoing the most profound demographic change in the country's history so that in a few decades, if not sooner, persons identified (and identifying themselves) as white and tracing their ancestry to Europe will have become part of the nation's racial and ethnic plurality, no longer its numerically dominant racial group."[2]

But even more to our point, the level of wealth within that demographic is changing. Overall, Black wealth increased by 321 percent in the years between 1989 and 2002.[3] Much of this is due to what sociologist Melvin Oliver calls the "social capital"—"living in nice neighborhoods with good schools for their kids"—acquired by the generations that were part of and immediately followed the civil rights era.[4]

This social capital was largely earned, not through success in the fields of sports or entertainment, but through investment in education and a lot of long, steady climbs up corporate ladders. If your personal experience and awareness of Black wealth is limited, we also remind you that this is in part perhaps because AAAs are a discerning and understated segment.

"To be fair, to really identify the wealthiest black Americans would take a lot of digging—after all, truly wealthy black people, (including many of the corporate CEOs, Wall Street executives and owners of *Black Enterprise* 100s' companies featured in *Black Enterprise*) are not eager to draw attention to their wealth. They're quite happy to let [rappers and athletes] get all the attention—and aggravation—that comes when people realize you're earning big money."[5]

Acknowledging that the niche audience of AAAs has grown and will continue to grow is mandatory for success in the contemporary luxury marketplace.

Finally, to attract and accommodate this growing niche, your marketing strategy has to become attuned to how this group wants to be courted. "Affluent blacks are major consumers of luxury brand-name goods."[6] That statement is certainly true. But it's also true that:

> Targeting black consumers is subtly different than marketing to a mainstream audience. Advertising requires a different approach. Inserting ethnic faces into traditional advertising campaigns is no longer effective. Communications require an elevated level of cultural awareness and under the radar media platforms. Otherwise, there is a risk of reinforcing ethnic stereotypes and alienating black consumers.[7]

So, how can you acquire that elevated level of cultural awareness? How do you develop and implement the strategies that will give your goods and services entry into this segment and make them brand-loyal consumers? What media and marketing platforms will you now consider partnering with? How will you be more inclusive? How will you launch an initiative or pilot program and when? Who will help you get started?

OUT OF SIGHT, OUT OF MIND: CONSISTENCY IS KEY

Harley-Davidson isn't a brand one automatically associates with the AAA audience. In fact, Tracy Ulrich, the marketing director of two Harley-Davidson dealerships in Toledo, Ohio, freely admits that her audience had been in the past made up of "mostly white males forty and up—from laborers to doctors and lawyers. But what [we] decided was to reach out to women and to ethnic groups."

Reaching out to a more diverse market was a smart move in a city where nearly a quarter of the population is African American. "We knew there was a large group of potential buyers that we were missing and we wanted to reach them," said Ulrich. "I think we had gotten stuck in that traditional advertising mode, and needed to look beyond it. Between myself, our owner, and our general manager, we realized that there's a lot of other groups out there that we want to get and share our lifestyles with. We want everyone to feel comfortable and welcome when they come into the store. The entire Harley-Davidson company has taken this philosophy in the past year or two—opening up to different groups. It trickles down to all of the dealerships."

To make this company-wide philosophy work at the local dealer level, Ulrich first formed a marketing partnership with the Northwestern Ohio Sickle Cell Foundation, advice she received from an African American consultant. The Sickle Cell Foundation is dedicated to raising money and awareness for a blood disorder that's particularly prevalent among African Americans. "Sickle Cell is an organization with a strong presence in our areas, but we also think it's a very good cause, and a hopeful one," Ulrich told us, explaining that the dealership had organized a Labor Day fundraising ride on behalf of the charity that was such a success they started looking to expand such rides to charities focusing on women's and Hispanic issues.

Yet, even while Ulrich looked for new opportunities to expose women and other ethnic groups to the Harley-Davidson lifestyle, she didn't stop marketing to the AAA community after that first successful ride. In January 2007, her dealerships sponsored a first-ever Martin Luther King Day promotion which targeted the area's affluent African American population by donating 5 percent of ancillary sales proceeds to the Sickle Cell Foundation.

With the direction of a consultant seasoned about this consumer, she promoted the MLK Day sale by placing spots on local urban contemporary stations. Ulrich said the response to these spots "convinced us that [urban contemporary radio] is a valuable venue, and we're ready to make buys on these stations a regular part of our standard marketing budget." She began to work with one of the stations to develop the marketing relationship the dealership enjoyed with them—"things like setting up a tent with some of our bikes or providing leather jackets for them to raffle off at concerts."

The result of the MLK event was an increase in motorcycle sales, as well as in motor clothing and parts—no small uptick when you consider a typical piece of Harley-Davidson clothing sells for about $125, or that the price of an average bike ranges from $8,000 to $30,000.

Ulrich reported that the promotion helped to deliver post-event sales as well. "Yes, definitely. When we are having other in-store events and parties, I'm seeing people from the MLK event. These other events are always related to a charity too—we do two or three events a month from March to September, and at least one a month in slower months." And the dealerships have also expanded their outreach to the AAA audience specifically, to "a month-long Black History series of events, instead of just the January weekend MLK sale."

What lessons can we take away from Ulrich's success? The first and possibly most critical is that she used a seasoned consultant familiar with her target market as she planned her campaign. But what about the practical steps she took on the advice of her consultant? Ulrich's first important one was to use urban contemporary radio to advertise her dealerships' first AAA target event. Again, urban contemporary might not be the first format that springs to mind when as a marketer you think about placing a media buy for a product like Harley-Davidson. But it's worth noting that advertisers targeting all African American consumers spent $2.3 billion in the last year, and they spend more money on radio than on any other medium, according to Nielsen Monitor-Plus.[8] They spent $805 million on radio—or 35 percent of total spending—usually on formats such as urban, Black news/talk, gospel, and smooth jazz, writes *Mediaweek*.[9] Note that the urban category spans a variety of musical tastes; it's not exclusively hip-hop or rap.

Ulrich was open to trying new venues to increase her potential customer base, and she used this particular one in an intelligent way: pairing her radio buy with the invitation to a specific retail event. As she put it:

> "Marketers shouldn't be afraid of taking the necessary steps in reaching new audiences, through new marketing efforts and partnerships. Particularly at the retail level. There are many groups of consumers that are not being reached through traditional advertising techniques. We should do our homework in learning how to reach them in relevant ways. Part of which needs to be activating retail engagements."

Additionally, the retail event was tied to a charity that had meaning within the local AAA community—in this case the Sickle Cell Foundation. When you can partner a brand to benefit a non-profit that has significance within your target demographic, you have the advantage of being able to call on the charity for access to its mailing lists, and publicity in its internal publications. You also activate in your behalf the base of support the charity already enjoys.

The most critical part of what we can learn from Ulrich's promotional plan, however, is that she didn't simply sit back after one triumphant Labor Day ride and decide that she had now forever captured the AAA audience in her area. Instead, she immediately went on to plan her next AAA target event, and the one that would come after that.

FREQUENCY MEANS LOYALTY TO
THE AAA MARKET

We applaud Ulrich's follow-through because all too often marketers sit back and rest on their laurels after the subsequent uptick in sales from an initial overture. It's frustrating to see a client who makes a first effort at reaching the AAA audience, only to ignore what he has invested after seeing a positive result because he believes that this audience should now be his. "Sure, we had a little pop at first, and then it went off," one of our clients once said to us, shrugging, leaving us to explain that the "pop" wasn't sustained because the advertiser went away. He made no continuing attempt to keep his message in front of the AAA audience—no more spots, no more magazine ads, no more TV ads, no more events, not even Internet or e-mail—and this audience, in turn, didn't go out of their way to seek out his brand.

Sometimes a lack of follow-through is due to more immediately practical matters, such as budget shifts. Unfortunately, budget cuts are indeed the sort of thing that can sink a marketing strategy, because many marketers have an "all or nothing" attitude. But, while the reality may be that his budget doesn't enable the marketer to produce a lavish campaign that incorporates all eight of the components of a 360-degree plan 365 days a year, the right thing to do in this instance is not to simply go away. It is to continue to put a little bit of money against this audience—and to do it in a savvy way that both accommodates the new budget and allows him to continue to keep this audience engaged. For example, in 2008, a luxury car company that had been advertising with *Uptown* wasn't able to commit to the magazine at the same levels it had spent in 2007. We advised them to cut back if they had to, but not to cut out. Even one or two small events a year, we told them, could enable them to keep the relationship with the consumer that they had been working so hard to grow. Then, knowing that a consumer is several times more likely to buy a car after a test drive, we went a step further; we suggested incorporating diversity tactics into the company's evergreen Ride and Drive events. Specifically, we suggested that the invitation list simply be more inclusive, be sent to the magazine's mailing lists, and that the company should also make sure to send invitations using the lists of the other diversity programs they were involved in. It was a matter of not just giving in or walking away, but of finding a creative way to remain in touch with the audience and leveraging relationships with existing partners and media outlets. It was a matter of both marketing people and their media partners being open-minded and receptive to a plan of action that was less traditional but, if done right, maintained audience share the company had already attracted and sustained its attention and affinity even during a leaner budget year. This is the "inclusion" part of a "diversity and inclusion" strategy.

As Ulrich demonstrates, in order to build allegiance to your brand, you have got to show your customers your appreciation for their loyalty on a continual basis. Otherwise, it comes off as inauthentic.

LOOK UNDER YOUR NOSE: RECOGNIZE AND ACKNOWLEDGE OPPORTUNITY

Another smart aspect of Ulrich's marketing plan was how astute she was in simply looking under her nose to find a whole new audience for her products. She knew that nearly a quarter of her city's population was African American, but she wasn't seeing a proportional number of Blacks walk into her dealerships. So she reached out to them and changed that. It helped immeasurably that Ulrich's efforts had the solid support of upper management.

Aston Martin, the makers of legendary British performance cars, had a similar awakening when they looked at their sales statistics and realized who their customers actually were. A sizable percent of Aston Martin cars were being sold in the United States, and of all those Aston Martins being sold, a substantial amount was being purchased by AAAs. Talk about an Aha! moment. With this knowledge, Aston Martin came to *Uptown*. They wanted to create an integrated program utilizing the *Uptown* brand and its relationship with this audience. It was a good way to reinforce the purchase decision, reach new audiences, and show the buyer that this company is willing to come to you where you play. Again, it's about lifestyle marketing.

As we've said several times, it's important to meet the AAA segment in their own playground. Ulrich, for example, sought out the radio stations with the largest Black listener demographic to advertise the events the Harley-Davidson dealerships were sponsoring to target this segment. She didn't double down on the standard buys she might have made at the local top-forty or country music stations, hoping to catch the ear of a stray AAA listener; she went right to the sources that could most effectively deliver the specific consumer she was looking to attract. She went where she knew she would find the listener density of her target segment. Going where they would find the reader density of their desired audience is exactly what Aston Martin was doing by coming to *Uptown*.

Aston Martin's signature event was dinner at a high-end restaurant. This could have been a general market event, but by reaching out to *Uptown*'s AAA audience when creating its guest list, the brand affiliated itself with something more than just the magazine's readership. *Uptown* is a brand itself, and one that already has an affinity with this audience. Its imprimatur lets this audience know that the company is okay, that it has the people's best interests at heart, and one is well advised to do business with it. It's a screening process of sorts. In the way that an investor may be more inclined to put her money into the growth of a company that reflects her political and/or social values (say, one with a good record of promoting women or African Americans to executive positions), a brand with the *Uptown* seal of approval has a built-in advantage with this audience as a brand that recognizes and respects its needs and values. In this case, Aston Martin used *Uptown*'s VIP list to create their guest list, pulling from the magazine's subscriber base the tastemakers that it had honed from events held in the past. As ambassadors of this market, *Uptown* has a keen sense of who these tastemakers are and is able to leverage its relationship with this audience.

Aston Martin arranged the details of the dinner, which included a display of vehicles and presentations by the CMO and other company representatives. At the end of the day, two cars were sold as a direct result of this dinner.

One dinner. The sale of two $175,000 cars. You can do the math, but Aston Martin was very pleased. The table below from R.L. Polk shows that even in a down economy, this brand, and their attention to this market has shown positive results. When the economy does turn around, Aston Martin will be positioned to experience significant growth from their efforts.

Table 5.1   African American Share of Aston Martin

	2007 CYE	2008 CYE	2009 CYTD July 09
Aston Martin New Vehicle Sales	1,487	974	502
African American Aston Martin Sales	41	41	8
African American Share	2.76%	4.21%	1.59%

Note: Although Aston Martin's personal new vehicle registrations were down 34.5% between 2007 and 2008, the number of Aston Martin vehicles purchased by African Americans remained steady. Thus the African American share of Aston Martin personal new vehicle registrations improved by nearly 1.5% going from 2.76% in 2007 to 4.21% in 2008.
Source: (Based on Personal Registrations Only—Does not include Fleet and Commerical Sales)

The larger point is that to build a business, you need to build your invitation list. Remember that reach and frequency are critical, and this includes building your invitation list. There are two ways of doing this: organically, through the acquisition of lists, or virally, through the guests who have already attended an event.

And this should not be limited to big, once-a-year events or product launches. Challenge your agencies and planners to produce strategic events that expand the universe of prospects. Don't confine yourself to the same "100" people whose photos always seem to make the local socialite and media pages of the newspaper. The iconic French brand, discussed in chapter three, expanded their prospect list by sharing their partner's supporter database; there are charities in every city that would be eager to offer their top supporters an evening of private shopping in exchange for a portion of the profits—in this case the significant sum of $8,500. Indeed, consider creating a similar signature event that your brand can roll out as a series in different cities, states, and even countries. Again, one-off programs are costly and ineffective if you want to build affinity, top-of-mind consideration, and word-of-mouth marketing. Two keys to effective marketing are frequency and reach, and a well thought out pilot program that can be scaled up and out can help you to accomplish this goal.

As we mentioned above, another way to expand your invitation list is virally—simply ask each invitee to bring a guest (a "plus one"). It is a strategy that can work well for shopping and music events. Giving a guest the opportunity to bring a guest herself increases not only your event attendance, but the invitee's "insider" status as well. And we all know it is simply human nature that, when we are insiders, our dedication to the group—or, in this case, the brand—becomes more constant. And the next event, the same people shouldn't be on the guest list but the original invitee should bring someone new. This builds your reach and, of course, word of mouth.

USING THE GRASS ROOTS

There is power in numbers—and there are many service and social organizations with affluent African American memberships. These organizations are not only rich in number; they also represent the optimal way to target the market by tying together all the components important to the AAA: heritage, connectivity, and like-minded peers. By partnering with these groups, in lieu of traditional media outlets such as TV and print, you can start building relationships that will put your brand at the top of mind—and do so in a way that is cost effective and authentic. Not to mention the word of mouth that comes with it.

We don't mean to suggest that affiliation with meaningful cultural and other charitable organizations, and/or with Black media, is the only means of building significant marketing partnerships within the AAA community. But it is a vastly underused avenue. This may be because, unfortunately, many long-established organizations within the African American community are still under the radar of the general media and what is still the general marketing sector. If, for example, we named The Marathon Club, Alpha Phi Alpha fraternity, Mocha Moms, Kappa Alpha Psi, or Jack and Jill, do you think a general marketer would know what these groups are, whom they serve, or even how extensive their membership rosters are? It's unlikely. So we're going to tell you a little bit about them.

This is certainly not an exhaustive list of the charitable, cultural, professional, or fraternal and sorority organizations with which many AAAs are tightly affiliated. But each of these groups represents an opportunity to reach into the core of the AAA segment and promote goodwill—and thus brand loyalty—among these keen consumers.

PLACES TO FIND PROSPECTS: CHARITABLE, SOCIAL, AND CULTURAL ARTS ORGANIZATIONS

THE LINKS, INCORPORATED

The Links, Incorporated is an international, not-for-profit corporation that was established in 1946. Its membership consists of 12,000 professional women of color in 270 chapters located in forty-two states, the District of Columbia, and the Commonwealth of the Bahamas. It is one of the nation's oldest and largest volunteer service organizations composed of extraordinary women committed to enriching, sustaining, and ensuring the culture and economic survival of African Americans and other persons of African ancestry.

Links members are influential decision makers and opinion leaders—most are philanthropists, college presidents, judges, doctors, bankers, lawyers, executives, or the wives of well-known public figures. The Links, Inc. has attracted many distinguished women who are individual achievers and have made a difference in their communities and the world. They are not only business and civic leaders, but role models, mentors, activists, and volunteers who work toward a common vision by engaging like-minded organizations and individuals for partnership. Links members contribute more than 500,000 documented hours of community service annually to strengthen their communities and enhance the nation. The organization is the recipient of awards from the United Nations Association of New York and the Leon H. Sullivan Foundation for its premier programs.

Opportunities to partner with The Links and attract the attention and affections of 12,000 powerful women include their numerous annual social activities that encompass debutante cotillions, fashion show luncheons, auctions, and balls.

100 BLACK MEN OF AMERICA, INC.

Founded in New York City in 1963, 100 Black Men began with nine chapters as a national alliance of leading African American men of business, industry, public affairs, and government dedicated to improving the quality of life for African Americans, particularly African American youth. Today, under the direction of Albert E. Dotson, Jr., the organization has 110 chapters, and growing, in the United States, England, and the Caribbean. Members represent a myriad of professions including corporate executives, physicians, attorneys, entrepreneurs, entertainers, elected officials, professional athletes, educators, and numerous others that have come together to form an international coalition focused on creating educational opportunities, promoting economic empowerment, addressing health disparities, and creating positive, nurturing mentoring relationships. The men of the 100 are committed as both individuals and as an organization to increase engagement with youth, and to influence others to become mentors, literally changing the future for a young person. This change reverberates exponentially throughout African American communities, our nation, and the world as 100 Black Men of America seeks to serve as a beacon of leadership; utilizing their diverse talents to create environments where African American children are motivated to achieve and are empowered to become self-sufficient shareholders in the economic and social fabric of the communities they serve.

Your partnership opportunity with your local chapter of 100 Black Men might be patterned upon the one that Aetna, one of the nation's leading health care benefits companies, enjoys with the organization's chapter in Atlanta, Georgia—an area of the country where Aetna has directed much of its support:

> "Aetna...has a long-standing relationship with the 100 Black Men of Atlanta, working with the organization on a wide variety of health-related programs. Aetna and the Aetna Foundation have given $230,000 to the 100 Black Men of Atlanta over the past three years, targeting funds to efforts to improve awareness of childhood obesity and diabetes, both of which affect the African American community with high frequency. Recently, Aetna announced that it will provide $10,000 to create a memorial fund in honor of the late Terrell L. Slayton Jr., past chairman of the 100 Black Men of Atlanta. The funds are part of an $80,000 Aetna Foundation grant."[10]

Though Aetna does not represent a luxury product, what luxury marketers can take away from its partnership is a good example of how to forge a deep and long-term relationship with an organization such as 100 Black Men of America—and what such constancy can achieve in terms of developing good will with the thousands of African American professionals who are its members. HSBC is another example of such a long-term relationship, but with a cultural institution: the Evidence Dance Company.

BOULÉ

Established in 1904, the Sigma Pi Phi fraternity, also known as the Boulé, was the first Greek-letter fraternity to be founded by African American men. Significantly, unlike the other African American Greek-letter organizations, its members already have received college and professional degrees at the time of their induction. Its membership consists of some of the most accomplished, affluent, and influential Black leaders dedicated to making lasting contributions to their communities, our society, and the world. The Boulé Foundation is a strong financial entity for providing the aid and assistance that is central to their mission.

There are 119 chapters of Boulé throughout the United States and the Caribbean. Its members have a median annual income of $250,000, and an average net personal worth of $2 million. Partnership opportunities include participation in the group's annual regional conferences and a national conference that convenes every two years.

THE GIRL FRIENDS, INC.

There is an organization known as The Girl Friends, Inc. Founded in 1927 in New York City, the group is composed of individuals from well-respected, Black upper-class families and sponsors many philanthropic and cultural activities that include raising money for charities. The group also sponsors social activities for its members. It includes about forty chapters in major American cities such as Chicago, Los Angeles, and Atlanta, with approximately 1,300 members nationwide. Membership is by invitation only and women are admitted only after being nominated by at least two existing members and then approved by at least two-thirds of that particular chapter.

THE MOCHA MOMS, INC.

There is the aforementioned Mocha Moms, Inc.; a support group for mothers of color who have chosen not to work full-time outside of the home in order to devote more time to their families and communities. Mocha Moms serves as an advocate for those mothers and encourages the spirit of community activism within its membership. The organization began humbly enough, with the publication of a newsletter called *Mocha Moms*. The goal of the newsletter's founders was to:

> "connect at-home mothers of color with each other. The newsletter was intended to encourage these mothers to feel good about their choice as well as to provide information to help them be the best and most important influence in their children's lives. It was distributed to over one hundred moms across the country in the spring of 1997. During the summer of 1997, moms in Prince George's County, Maryland decided to form a support group called Mocha Moms, Inc. There are now over one hundred chapters of Mocha Moms, Inc. throughout the United States and the organization is continually growing and evolving to meet the needs of our moms, their families, and the communities they live in."[11]

In terms of charitable, social, and cultural arts groups there are also the National Black Arts Festival, the Twenty-First Century Foundation, and Jack and Jill, a group for which your brand might take the opportunity to sponsor a teen volunteer day. As we said, we're just giving you a taste here of the myriad institutions that are comprised of the top-tier of the Black upper class, and of the opportunities to reach the hearts—and pocketbooks—of their members.

PROFESSIONAL MEMBER-BASED ORGANIZATIONS

THE EXECUTIVE LEADERSHIP COUNCIL

The Executive Leadership Council is preeminent among African American member-based organizations. It was formed to recognize the strengths, successes, contributions, and impact of African American corporate business leaders. Its mission: to build an inclusive business leadership pipeline and to develop African

American corporate leaders—one student and one executive at a time. This leadership network is guided by a bold and historic vision of inclusion, which is the leadership legacy of African Americans—whether in business, education or the community. With more than 390 members, one-third of them women, the Executive Leadership Council is the nation's premier leadership organization comprised of the most senior African American corporate executives, pro-diversity CEOs, scholars, and business leaders in Fortune 500 companies. Its membership represents well over 250 major corporations.

What are some of the ways you can connect your brand with these top level executives? Well, your brand could be one of the sponsors of the Executive Leadership Council's Annual Recognition Dinner and Awards Ceremony. You could become involved with the Corporate Board Development Program, which works with the National Association of Corporate Directors (NACD) and the Kellogg School of Management at Northwestern University to provide Council members with training and research about the responsibilities of board directors, as well as the latest in board leadership topics. Since its inception, more than 600 Council members have taken part in the Board's development programs, which have enjoyed sponsorship by the Altria Group since 2002.

Other sponsorship opportunities include the Annual CEO Diversity Summit, which brings member company CEOs and Council members together in a day-long learning environment and in which more than 200 Fortune 500 CEOs have participated since its inception in 2001 by founding sponsor GE. There is also the Black Women's Leadership Summit and Black Women on Wall Street, a two-day leadership development symposium developed by Merrill Lynch at which, in 2007, Citigroup sponsored an additional half-day session for women in mid-level management positions. Spring & Winter General Membership Meetings are held nationwide. The Annual Mid-Level Managers' Symposium (MLMS) is supported by founding and continuing lead sponsor the PepsiCo

Foundation, and has recently been cosponsored by the Principal Financial Group and ING.

This is just a small sampling of the opportunities available for partnering with one organization—opportunities that companies such as American Express, AT&T, Hewlett Packard (HP), JP Morgan Chase, Key Bank Corporation, Moody's Investors Services, United Parcel Service (UPS) Foundation, New York Life Insurance Company, Sprint Nextel Corporation, Daimler Chrysler, BMW, IBM, Oracle, and Entergy have taken advantage of in the past.

THE MARATHON CLUB

The Marathon Club was founded by the National Association of Investment Companies (NAIC), a financial industry association for private equity firms that invest in an ethnically diverse marketplace, and the New American Alliance (NAA), an organization of American Latino business leaders. Indeed, the Executive Leadership Council joined NAIC and NAA in 2004.

Together, as the Marathon Club, they are focused on increasing the availability and investment of private equity capital into enterprises that have significant minority ownership and management participation. Additionally, the Marathon Club works with other organizations to develop research and public policies that will increase the success of minority business enterprises. Marketing partnership opportunities with the Marathon Club include the program Annual Corporate Partners; customizable sponsorship packages that allow corporate partners to connect with highly affluent and prominent business leaders across a variety of disciplines and cultural backgrounds. Also available is the MC2 online platform, which offers exclusive visibility for advertisers in a proprietary networking environment. Corporate partners in this online venue achieve high-level exposure and branding opportunities in an arena that TMC members associate with collaboration and value creation.

There are also several events associated with the Marathon Club. The club's signature event is the DealMakers Summit, a national forum where hundreds of top professionals converge to create value and harness the extraordinary opportunities available in this market. Regional networking receptions are also held regularly in cities across the country. These invitation-only events provide business owners, investors, Fortune 500 executives, and professional service providers the opportunity to connect informally with like-minded professionals across various industries and cultures. A significant opportunity for marketers hoping to partner with this exclusive group is the Marathon Club's Wealth Creation Seminar Series. These seminars are intimate gatherings at which successful multi-unit franchise owners and hotel developers, among others, share their experiences about how participants may access and benefit from these paths to significant asset building and the role private equity can play in this process.

THE NATIONAL ASSOCIATION OF INVESTMENT COMPANIES

Founded in 1971, the National Association of Investment Companies (NAIC) is *the* financial industry association for private equity firms that invest in an ethnically diverse marketplace. Since NAIC's inception, its member firms have invested in well over 20,000 ethnically diverse businesses. Utilizing its collective expertise, NAIC has helped to create a marketplace full of opportunities for growth, success, new jobs, and an expanded tax base. Today, NAIC member firms manage more than $10 billion in capital, and NAIC members cover the full spectrum of private equity investment activity, including early stage venture, later stage venture, expansion, buyout, mezzanine, distressed, and secondary fund investments.

Marketing partnership opportunities abound here as well. NAIC hosts several educational forums and professional conferences each year, including an annual convention, the Minority Manager Consortium.

THE BLACK RETAIL ACTION GROUP

Black Retail Action Group (BRAG) is a not-for-profit organization dedicated to the inclusion of African Americans and all people of color in retail and related industries. You might reach their 200 members by participating in the group's annual scholarship and awards dinner gala, an event that recognizes outstanding achievements made by persons in retail and related industries, as well as other business and industry sectors, the arts, and education. You might also get involved in their Executive Development Series, a unique interactive series of workshops designed to help individuals at all stages of their career identify and devise action plans to achieve a breakthrough to success.

THE NATIONAL ASSOCIATION OF BLACK ACCOUNTANTS, INCORPORATED

Since 1969, the National Association of Black Accountants (NABA), has been the leader in expanding the influence of minority professionals in the fields of accounting and finance. NABA is comprised of professional and student members who strive to maintain the highest standards of professionalism in their careers and academia, all while advancing the group's mission to develop and inspire future leaders who will shape tomorrow's accounting and finance professions. It encompasses over 200 professional and student chapters across the country and has over 8,000 members.

THE NATIONAL SOCIETY OF BLACK ENGINEERS

The National Society of Black Engineers, founded in 1975, is the world's premier organization serving Black engineers, scientists, technologists, and mathematicians. The mission of its 32,000-plus members is "to increase the number of culturally responsible Black engineers who excel academically, succeed professionally, and positively impact the community." NSBE is comprised of more than

350 chapters on college and university campuses, with seventy-five Alumni Extension chapters and seventy-five pre-college chapters nationwide.

The National Association of Black Journalists, the Association of Black Psychologists, the Organization of Black Screenwriters, the Organization of Black Airline Pilots, the National Society of Black Physicists, the Black Ivy Alumni League, the National Black MBA Association, the National Minority Golf Foundation, and the National Black McDonald's Operators Association—the list goes on to include over sixty other outlets, organizations, blogs, and more that speak directly to the affluent new niche you are trying to reach. Do you have vacations in Gstaad to sell? Perhaps you want to reach out then to the Black Brotherhood of Skiers. Do you have a new, white-table cloth restaurant on the order of the Obamas' favored health food place in New York, Blue Hill, to promote? Perhaps you should reach out to www.blackdoctors.org to build your client base. Are you planning a retail event for your fine goods? Maybe the members of the Council of Urban Professionals should get invitations.

FRATERNITIES AND SORORITIES

Historically, Black Greek organizations remain important to their members long after they have left their respective colleges and universities. And each of these groups has alumni chapters around the country. In his book, *Our Kind of People: Inside America's Black Upper Class*, Lawrence Otis Graham says that these sororities and fraternities "are a lasting identity, a circle of lifetime friends, a base for future political and civic activism." They are also a valuable resource for marketers with the foresight to recognize that, in addition to appealing to lifelong members, these organizations represent the opportunity to attract young, aspirational college students—the younger demographic who will fill the prospective pipeline for luxury brands.

Beyond the groups we've already listed, there are 105 historical Black colleges and universities in the United States and the Caribbean. And there are a total of nine historically Black sororities and fraternities that make up the National Pan-Hellenic Council, sometimes referred to as the "divine nine." The organizations include Alpha Phi Alpha Fraternity, Inc. (1906); Alpha Kappa Alpha Sorority, Inc. (1908); Kappa Alpha Psi Fraternity, Inc. (1911); Omega Psi Phi Fraternity, Inc. (1911); Delta Sigma Theta Sorority, Inc. (1913); Phi Beta Sigma Fraternity, Inc, (1914); Zeta Phi Beta, Inc. (1920); Sigma Gamma Rho, Inc. (1922); and Iota Phi Theta, Inc. (1963).

These lists are an important piece of the 360-degree plan. They show the potential that awaits anyone who seeks out long-term affiliations with one or more of these organizations. Here are a few success stories of brands that have already done this.

Nations Bank Corporation, based in Atlanta, Georgia, is a role model—an early adaptor—to this opportunity. As far back as 1996, Nations Bank decided to target wealthy Black professionals because they were viewed as a viable group of prospects.

Private banks have historically sought out niche markets. But in 1996, no other banks were looking to this particular segment. It likely wasn't a fluke that this strategy got started in Georgia. The Atlanta metropolitan area, specifically, stands out as an incubator for Black entrepreneurship. The city was also a focal point in the rise of African American political power. Following the election of civil rights veterans Andrew Young and John Lewis to Congress, and beginning with the election of Maynard Jackson as mayor in 1974, all of the city's mayors have since been Black. Today, in Atlanta, the Black population numbers 1,502,745, with an AAA population of approximately 82,500.[12] Possibly the only thing surprising about Nations Bank's outreach to the Atlanta AAA community was that its managers and marketers were able to recognize and act on the changing demographics at the earliest evidence of the groundswell.

At the time, Shedrick L. Barber, the national coordinator for the bank's Professional African American Market Development department, said, "We found that this group has an annual purchasing power of $427 billion, and it's growing...We found a new customer in our back yard."[13] The bank's research showed that in 1996 there were more than one million Black households in which one member made at least $50,000 a year, and "what matters even more to Nations Bank, seventy-five percent of those [were] in the Southeast."[14]

The initial ROI after the launch of this new strategy showed that Nations Bank had "booked $148 million in 37 deals," and had an additional "$500 million in business in [the] pipeline, including mortgage loans, franchise financing, securities and other products."[15] It managed these amazing numbers marketing to Black professional associations. Nations Bank was the title sponsor of the first Black Enterprise Magazine Entrepreneurs Conference that was attended by more than 500 Black business owners. One of the bank's first loans was a $25 million deal with Omega Psi Phi. Barber entered into a marketing partnership with the Black Automobile Dealers Association and the Meharry Medical College, a top school for Black doctors. "The people I'm talking to could get their deals done anywhere," Barber said.[16] But, in this case, Barber sought them out—and they did their lucrative deals with Nations Bank. Why? Because Nations Bank acknowledged them, and they reciprocated.

More recent examples of the brands that are reaching out to the AAA segment can be found on the sponsor list for the annual Black Enterprise/Pepsi Golf and Tennis Challenge. In 2009, this sporting event celebrated its sixteenth anniversary; this event has become a Labor Day tradition for nearly 700 affluent African Americans. Among the brands welcoming the guests to its event were platinum sponsors American Express, American Airlines, and Polo Ralph Lauren. Another example would be Northern Trust and American

Express' sponsorship of "In the Hands of African American Collectors: The Personal Treasures of Bernard and Shirley Kinsey." This art collection travels around the country and includes works of art by important African American artists such as Romare Bearden, Elizabeth Catlett, Sam Gilliam, William H. Johnson, Jacob Lawrence, and Henry O. Tanner, as well as historical documents and artifacts of Benjamin Banneker, Frederick Douglass, Harriet Ann Jacobs, Alain Locke, Phillis Wheatley, and Malcolm X. When viewed as a whole, the ninety-plus objects reveal important aspects of American history and culture.

How many ways are there for your brand to participate in and benefit from marketing partnerships with these sorts of organizations? Let us count them: signage at events, program advertisements, retail booths, sharing mailing, and member lists. While we want to affirm that a 360-degree integrated marketing approach is the most efficient avenue, perhaps your brand's participation in such a program could start as simply as providing a gift for each place setting at a member convention, private dinner, or hosting a kick-off event at your retail store.

If you are going to offer a gift at an event as part of your marketing partnership, we would prefer to see a brand with an exclusive sponsorship in this area; one high-end brand that offers a gift card or certificate of substance, presented in a beautiful envelope. The clear advantage of a gift card is that it will bring this wealthy consumer into your online store or bricks-and-mortar retail establishment. It also lends itself to creating and implementing a Prospect Relationship Management system or "PRM." But if you are more attuned to offering an item, make it something of significance. Brands will sometimes see such giving as too lavish—that is, too expensive—and when they do we remind them of three important factors. One, events at which such gifts are presented are usually smaller ones—gatherings of forty, fifty, maybe sixty, elite potential customers. Two, the customer drawn into your retail store because he or she has a gift card to spend, or because they like a gift so much they want to see what else your brand has to offer, often spends tenfold over the value of the gift. And, three—likely the most apparent point, though frequently missed—you are selling a lavish luxury brand, which is better done when you treat your customers, and potential customers, lavishly in return. And this still beats an ad in the *Wall Street Journal* because it directly targets the prospects, gets them to your store, and does so for a fraction of the cost.

But even as we're taking advantage of the obvious opportunities, let's not overlook the hidden extras of partnering with these organizations. These organizations are not stand-alone local units. They are, for the most part, chapters of nationwide institutions, and therefore have a broad constituency that reaches beyond any one local area. They are all a part of the vast AAA network we have already talked about. In sponsoring one luncheon for, say, The Links, notice how your brand's hospitality will reach the ears of that organization's members coast to coast, frequently as a news item in the group's internal newsletters. In speaking directly to forty, fifty, or even sixty, of the organization's members at a small, elite function, you speak indirectly to thousands more—and within a segment where positive word of mouth is key. Plus, you may be creating relevance and relationships where there are none and building inroads that can be leveraged in the future while combating the advances of better financed rivals. Remember, if you are loyal to this consumer, it'll be hard for the competition to infiltrate.

BORROWING EQUITY—PARTICIPATING IN ESTABLISHED EVENTS

When Sony wanted to explore the untapped affluent ethnic consumer market, the company's corporate marketing manager, Kadesha Boyer, contacted Diversity Affluence. "With a reputation as a leader and innovator, this opportunity appeared to be worthwhile to Sony because we want to stand apart from our competitors," Boyer said. She explained the company management's commitment to diversity this way, "We operate in a global business world and in order to compete, our clientele must reflect that."

Of the several different sponsorship ideas Diversity Affluence presented to Boyer for Sony's pilot diversity program, Boyer chose to have the company participate in two of the venues suggested. "The other events were on smaller scales—a few intimate dinners with maybe thirty to forty people—whereas the events we selected were much larger, which appealed to us." In other words, rather than create an event around the sponsorship of the Sony brand, Boyer opted for Sony to be a presence at functions that were already established brands themselves.

One of the events was the American Latino Media Arts (ALMA) Awards where Sony showcased its OLED (organic light emitting diode) television. "The screen size is eleven inches diagonal and features a 3mm thin panel. The picture quality is amazing, offering a contrast ration of 1,000,000:1," Boyer told us. "This product costs $2,500. You really have to experience it, which is why we liked the idea of these events. Consumers had the opportunity to interact with the product and ask us questions."

At the ALMA Awards, Sony sponsored an exclusive lounge area where they placed five OLED TVs. When the partygoers entered the lounge, they were offered champagne and strawberries. As they enjoyed their refreshments, they could touch the televisions, and ask the on-hand product specialist about them. Sony also offered a raffle, but in order to participate the entrant had to fill out a data capture sheet to help the company build a prospect database and receive insightful feedback about the brand. Sony was able to add approximately 250 names to their existing database through the two events Diversity Affluence suggested—and actually sold one of those high-end televisions on-site. Guests who filled out the forms received valuable Sony gift bags.

It might seem unconventional to you that a brand as well-known as Sony is taking this marketing route, so we'll let Boyer have the last word on that:

> [Sony has] brand awareness across the board, sure. From the events, we can tell that everyone knows and loves the brand. Everyone understands—even if we haven't directly targeted them—that we're known for quality and innovation. That's not the issue. The issue is whether or not we are connecting with and relating to them. Some forms that prospects filled out enlightened us to the fact that we aren't always reaching who we think we are, and our goal is to change that. The reality is that mass marketing is kind of old school. It may be more cost effective, but it's not as efficient, especially for higher-end consumers. Every consumer is different. It's not fair to lump them into the same bucket—their needs must be catered to.
>
> Sony is aware that its brand appeals to a diverse array of consumers with unique experiences and perspectives. For this very reason, mass marketing—while cost effective—is not the way we address our diverse audience. We opt for a more direct approach, catered to addressing each unique group. There are opportunities in the ethnic-affluent market. Sony strives to be a leader in this market and encourages others to follow suit.

Lexus is another brand that recognizes how imperative it is to cater to its clients. In fact, this top luxury automaker credits its decades of success on building vehicles in response to the needs of its customers. In the case of affluent African Americans, Mary Jane Kroll, Senior Advertising Administrator of the Multicultural Division of Lexus, saw and acted on the connection with AAAs to the company through its line of luxury hybrid vehicles. Research indicated to Kroll that Black people are more likely to feel personally affected by negative environmental impact, which results in a heightened interest in environmental issues. According to Kroll,

> "If you are a person who is a minority or a person of color, you're more likely to live near polluted places. You're more likely to have power plants where you live. We've participated in the TED[17] conference the past several years, and there was this speaker by the

name of Majora Carter and she founded an organization called The Sustainable South Bronx. She gave a talk that sort of lifted a veil of information that I don't think people had before that time. She was a lifelong South Bronx resident and saw that she had been personally living near these wasteland parks that had been built over or neglected. There's a strong sense of community among African Americans where they want to help each other and they really respect companies that step in and offer resources. Obviously there's been a [green] groundswell in the general market as well, but there's [also] been an added layer in the African American community. We've seen increases in awareness and concern among the African American community for both the environment and environmental products."[18]

It was at this point that Lexus recognized two crucial things. First, the company knew it was in a position to lead in both the AAA and luxury hybrid segment. As Kroll put it:

"Our hybrid products in the luxury realm obviously establish us as a leader because we don't have any competition, or at least not until recently with Cadillac SUV hybrids. So this has been an area where we've been able to establish our leadership and continue to build upon it as we introduce more products and as we build lifestyle programs around them to help people understand what the offering is. If Prius is everyone's idea of what hybrid means, it's actually a different story from a luxury standpoint. That's the starting point for [our] eco [program]."[19]

Lexus also recognized that it was critical for marketers not just to place ads in African American media, but also to get involved with the community and be an active contributor. To promote its four luxury hybrid models—the RX 450h, GS 450h, LS 600h L and the HS 250h—Lexus engaged the African American community. For its first outing, the company chose to be a presence—in a big way—at an existing event: a pre-Academy Awards event. This high-profile, celebrity-filled event had decided to "go green" for the first time. Lexus tapped into its eco-chic expertise by incorporating an exclusive roster of designer partners for its Lexus Hybrid Living program. A groundswell of buzz resulted and Lexus went on to be a part of the first-ever green issues of two African American media partners, and host additional evenings featuring hybrid vehicles at hip green venues. For Lexus, there is no question about the importance of marketing to African American consumers. The key is to continue to find the best ways to deliver an authentic brand experience and demonstrate how a Lexus vehicle can meet the diverse needs of its customers.

Lexus held a series of "listening lounges" that featured an upand-coming recording artist alongside one of their luxury hybrids. "In our Los Angeles lounge we did it at a green venue called The Smog Shop and we featured the RX 450h, the new generation of the current RX 400h. A few 400h's were sold directly as a result of the event," says Kroll.[20]

As to her take on niche versus general marketing, Kroll had this to say: "You can't rely on any single avenue to get your overall message across, but definitely with the African American community, you have to show that you want to invest in that community. Engagement events are a really effective way to do that because they help provide an authentic experience with the brand."[21]

David Nordstrom, Lexus Vice President of Marketing, had this to add:

> "The key to Lexus' success is that we always listen to our customers and provide them with the cars and services they want. Both Lexus and the African American community are concerned with the environment and are looking for ways to make less of an impact. Through special events that engage the community, as well as partnerships with key African American media, Lexus has been able to share the benefits of our four luxury hybrid vehicles in a meaningful way."

We've had our own authentic experience with Lexus, and you may well laugh when we begin to tell you about it. It was a terrible, wet, soggy evening. Lexus had invited *Uptown* readers to its dealership in Manhattan and, though we had initially expected seventy to a hundred people, we thought the weather was going to make it a bust. But about twenty people showed up. And over a six-month period, three cars were sold as a direct result of the event, one of them by a referral from someone who had attended—a perfect example of how word of mouth works in this community.

The thing people often miss is that these AAA consumers are so underserved. It's just a natural, human instinct to show gratitude to and interest in the people who go out of their way to speak to you. Now, these are the sort of people who are going to buy a luxury vehicle anyway, but Lexus was the car company that went out of their way to speak to them, so over 10 percent of those who came to that one Lexus event are now Lexus owners. Not a bad ROI. Imagine the additional revenue for the brand if the event was scaled up and scaled out?

VICTORIA'S NOT-SO-SECRET

There is another side to the story we've just told you. Sometimes you don't have to go out and find your customers; sometimes they tell you how to find *them*. Dwayne Ashley, the president and CEO of the Thurgood Marshall College Fund told us:

> "Following the civil rights movement of the 1960s, large numbers of African Americans entered the nation's colleges and universities, including HBCUs. So it's not surprising that during the past two generations, in particular, we have made notable gains in achieving higher incomes and affluence. Most of us, however, do not excel without the necessary family backing and community nurturing, so we are taught at an early age to be loyal to those who support that path to success. This cultural lesson, in my opinion, is very important for luxury marketers to understand about African Americans regardless of whether they are trying to reach Baby Boomers, or Generation X and Y."

In carving out a place for itself in the AAA segment, market-ers are well advised to not overlook the youth segment-within-a-segment of this audience. This audience that Gucci, Burberry, and Madison Avenue seem to be courting have a few hundred friends that they can spread the word to. They are the present and the future. Like the Obama campaign, savvy marketers can have an incredible impact on younger, emerging AAAs at an age when they are developing lifelong buying and shopping habits. Reaching the younger generation by being present in their homes, in their life-styles, and in their colleges will give you a pipeline well-populated with aspirational prospects. And, as you'll see with this segment, if you shortchange them, they'll tell you about it.

Victoria's Secret is not considered a luxury brand. It is, however, a very successful lingerie brand that enjoys the loyalty of many teen-age and college-aged women, and it is the subject of the interesting marketing phenomenon we want to tell you about. In July 2008, Victoria's Secret launched the Victoria's Secret "Pink" collection, a line of apparel designed for the young college-aged woman. The line incorporates various universities in their designs, among them Penn State, the University of Connecticut (UConn), UCLA (the University of California, Los Angeles), Michigan State University, and the University of Miami. But they had included not one histor-ically Black college or university.

Victoria's Secret consumer and Howard University sophomore, Amelia Reid, objected to the omission. When she saw a hoodie from the collection she wanted and went to buy it, she noticed the HBCU oversight and thought she'd try to set matters right by e-mailing Victoria's Secret and tipping them off to a market they were miss-ing. In response, Reid got a return e-mail from a customer service representative that she called "sugar coated."[22] Had this been sim-ply an issue of a school logo that had not been included, the mat-ter might not have progressed, but it wasn't. Benefits of a college or university partnering in the line with Victoria's Secret include paid marketing internships and scholarships for the students, as well as a portion of the sales being returned to the schools. So Reid decided to take things one step further. She launched the Facebook page, "HBCU Ladies Wear Victoria's Secret Pink Too." The page drew over 700 members. As Tina Wells, CEO and founder of Buzz Marketing Group said:

> "The African American consumer is brand loyal and will spend a lot of money on the brands they love. It was a huge oversight on their part not to have [included HBCUs initially]...I don't think they anticipated that what [Reid] did would get so many young women passionate about this. It goes to show the power of not just social networking but what happens when the beauty of the Internet puts communication in the hand of consumers to go direct to brands and saw, 'I want something to change.'"[23]

Change things did. The HBCU Pink Line launched at Florida A & M University (FAMU) in January, 2009. In the photo of the launch, an event that was highlighted in *Black Enterprise*, was the CEO of Victoria's Secret Pink brand, Richard Dent. It might surprise you to note that Dent is an African American and an HBCU alumni himself. We can guess only that sometimes even AAAs themselves have to be reminded of what a powerful market they constitute.

DIVERSITY MARKETING BY DEFAULT

One of the points we keep making is that putting a Black model in your print ad or a Black mannequin in your store window is no longer enough to attract the AAA buyer. At the same time, we don't want to give you the idea that there is no need at all to use Black models and spokespeople for your brand. As we've said earlier, it is always a positive experience to see oneself and one's lifestyle reflected in advertisements—especially if a company is urging one to spend money on its goods and services. Here, briefly, we do a call out for some of the brighter advertising lights for ways they have used people of color in their creative.

DIVERSITY MARKETING BY DESIGN

"After almost tripling their money spent on cruises from 1997 to present, African Americans are finally getting noticed by the travel industry," trumpeted Nicole Marie Richardson in an article for *Black Enterprise* in 2000.[24] The development was noteworthy, as African Americans had spent more than $2.1 billion annually since 1997 on three of the travel industry's major segments—cruise liners, airlines, and hotels—and now this segment was going to get a little recognition.[25] What is significant about the ads that resulted from this initiative is that they are models for the taste and understatement that appeals to this audience.

Let's take, for example, the print ad that shows an African American couple taking advantage of one of the cruise line's amenities—poolside couples massages. The couple is relaxed, attractively middle aged, and the pleasant smiles on their faces tell us that they are obviously enjoying the service. The cruise line's facilities are shown off splendidly; the scenery is breathtaking. So what is so out of the ordinary about this ad? *Nothing.* And that's just the point. There is no reference to urban or hip-hop in the photo. The subjects are anonymously midlife, not any sort of celebrities. No one is sporting dreadlocks. This could be a general market advertisement—and it will indeed likely serve to attract a whole range of consumers to the cruise line—because the couple highlighted just *happens* to be Black. At the end of the day, your best bet is to represent African Americans in your print ads as you would any of your more traditional luxury consumers. And that is just what Royal Caribbean has managed to do very well. Moreover, the ad was placed in the new African American online magazine www.theroot.com, which is owned by the *Washington Post.*

INFINITI

What the automobile maker Nissan sought to do with its new ad campaign for Infiniti was to "shape the Infiniti brand image among unaware African American consumers and showcase the shared values of unique, modern, distinct, and bold design."[26] The advertising agency, the True Agency, engaged in creating the campaign was charged with "[introducing] Infiniti and [illustrating] its distinct approach to design in a way that would captivate the imagination and interest of affluent African Americans."[27]

Before it embarked on any creative journey, however, the agency did its research on the AAA market and came up with these three "key brand attributes." One, their target audience took pride in Black culture and accomplishments. Two, their target audience was supporters of the arts. Three, this segment had a strong desire to be "respected, acknowledged, and appreciated."[28]

The True Agency executed an idea that appealed to all three attributes beautifully. "The big idea was to partner each vehicle with an African American cultural arbiter whose vision, creative process and expression reflect the vision and expression of each Infiniti. Infiniti in Black (IIB) would introduce visionary artists to African Americans—beginning the dialogue on Infiniti design through the artists' work."[29]

The five artists selected were: Kehinde Wiley, an emerging force in the art world who seeks to include African American men in works as a conscious correction for the dearth of a presence of African American males in traditional painting; Paul Miller, also known as "DJ Spooky that Subliminal Kid", an internationally renowned disc jockey, a professor at the European Graduate School in Switzerland, a composer who provided the score for the movie *Slam* that took prizes at both Cannes and the Sundance Film Festival, and author of the book *Rhythm Science*, which was published by MIT Press; Euzhan Palcy, a filmmaker who is noted for having been the first

African American woman to direct a mainstream Hollywood movie, *A Dry White Season*; Stephen Burks, an architect and industrial designer who has created cutting-edge designs for such major companies as Calvin Klein, Estee Lauder and Missoni; and Diedre Dawkins, a dancer and choreographer who "elevates and creates new experiences and emotions with every step, spin, jump and turn, elevating her performance on every level."[30]

What is it that these five artists, chosen by Nissan to represent their product, have in common? Simply that they are grown ups. These five women and men are all mature enough to have tested and pushed the boundaries of their art forms; they are serious artists who demand serious attention from a seriously moneyed segment such as AAAs.

DIVERSITY BY DESIGN VERSUS BY DEFAULT

Yves Saint Laurent was famously the first major designer to put a Black model on his runway. At the time of his death in 2008, at the age of seventy-one, supermodel Naomi Campbell paid tribute to the man who was also her personal champion: "My first *Vogue* cover ever was because of this man. Because when I said to him, 'Yves, they won't give me a French *Vogue* cover, they won't put a black girl on the cover' and he was like 'I'll take care of that,' and he did."[31]

If YSL paved the way in the 1970s for Beverly Johnson, Naomi Simms, Veronica Webb, Iman, and other top Black fashion models, and if Naomi Campbell, Tyra Banks, Claudia Mason, and Alek Wek were able to follow in their footsteps, posing on the runway and in print ads for Ralph Lauren, Vivienne Westwood, Valentino, Fendi, and Chanel, we think it takes nothing away from his groundbreaking innovations to say that it could have been a simply pragmatic move and still be a brilliant one: Black women spend $20 billion on apparel each year.[32]

In much the same way, if we look beyond the beauty of the new faces of Lancôme—Arlenis Sosa who hails from the Dominican Republic—and Estee Lauder—the Ethiopian model Liya Kebede, the first Black woman to represent the company in its fifty-seven-year history—we can see the decisions to use these women as one that is aesthetic as well as practical. African American women account for 19 percent of all cosmetics sales in the United States. Even more telling is that when Estee Lauder began featuring Kebede, they noticed an increase in sales of their darker foundation shades and had to expand on a darker palette of shades to meet the needs of the affluent women of color who were starting to flock to their counters. This is what we call diversity marketing by default, not design. Now imagine the significant increase in revenue if this message were to be pushed out through additional Black media, or a precise strategy was designed to reach this customer. These are facts and figures that marketers should look at closely and take to heart—there are still plenty of brands that could benefit from portraying women and men of color as users of their goods and services. For example, according to the NPD Group, a leading provider of consumer and retail information for a wide range of industries, African American women are not only more likely than any other group to use fragrances, but they wear fragrances far more frequently than the rest of the ethnic groups. Almost half of African American women indicate they use fragrances almost every day.

What should be central, however, to an understanding of how models like Kebede have been incorporated in the companies marketing strategies is, as Estee Lauder president Patrick Bousquet-Chavanne puts it, "The choice of Liya herself was first linked to her style and personality."[33] That is, Kebede will be in Estee Lauder's ads in publications such as *Essence*, but she will also be in *W* and *Vogue*; she didn't get her contract in order to appeal to only women of color, but to all women who want to be stylish.

IS THERE SUCH A THING AS LUXURY ONLINE?
REACHING NEW MARKETS IN NEW WAYS

The Internet, and all its myriad ways of reaching your customers and potential customers, is the final piece of the 360-degree puzzle. But if luxury is quickly becoming about experience over product, and brand awareness depends more on close customer relationships over flashy marketing stunts, is it possible for luxury brands to survive online? Isn't "online luxury" an oxymoron? This was the challenge faced by Denise Incandela, senior vice president for Saks Direct, the online and catalog division of the tony New York-based department store Saks Fifth Avenue.

"The Saks brand promise has always been to 'expertly deliver personalized style,'" said Incandela. "It was hard to imagine how to deliver on this promise and maintain the luxury experience online." Then Incandela decided to look at the Web differently: "We found out that most women are online at around 2 a.m.—after they have taken care of the kids, the house, the job, the pets. And we figured, what is more personal than providing customized shopping service on her schedule, when she's available?"

Incandela's team created an online shopping experience that features 24/7 live chat service with a knowledgeable personal shopper, editorial style advice, product suggestions to match choices that shoppers had already placed in their carts, large photos, and as much product detail as possible for each item. "We looked at it as the ultimate integrated shopping experience. In the actual store, it's very hard to move from department to department with the same sales associate so that you can check out once and buy the shoes with the dress and the accessories. Online we can do all that, and as long as we have a very well-designed site, the experience still feels very high-end."

Despite the challenges and unsettled current state of luxury retailing, the future of luxury is bright. That is, if marketers can create new ways of finding out where the money is in this new age of changing demographics, changing spending habits, and changing technologies. The case studies we've discussed in this chapter provide insight into the ways innovative marketers are responding to this era of change in effective and profitable ways—and we hope they provide inspiration to you as you seek to make a connection between your brand and new affluent segments. Our final case study, in the following chapter, will show you the steps we take when we create a total marketing plan for a new client—and we encourage you to use it as a template for creating your own cutting-edge marketing program.

STATING THE OBVIOUS: DON'T FORGET P.R.

In the effort to reach their customers, most companies use mainstream media, but never think to branch out into targeted outlets. Similarly, no marketer worth his or her salary is going to stage an event for a client and not put out a pre- or post-event press release to build a buzz. We want the press to come to the events we produce, and to cover them, and sometimes we will go to extraordinary lengths to get editorial about our event in the business section or a photo showcasing our client in the social pages. But when it comes to something as simple as sending out press releases for an event focusing on the AAA segment, many marketers are like party hosts sending out invitations to all the wrong addresses. They often fail to notify the 40,000-plus journalists and bloggers working for over 850 Black newspapers, magazines, television and radio stations that they have something going on that might be of interest to them.

There are, however, Black public relations services that can easily remedy this oversight and connect you and your brand to the network. BlackPRWire, for example, the first Black American newswire service, distributes around-the-clock client press releases, video and audio news releases, electronic video messages and electronic newsletters to African American, Caribbean and African newspapers, magazines, radio and television stations at the state, regional, and national level. UrbanAdserve is the leading online ad network for savvy urban publishers, marketers, and brands that want to connect with urban trendsetters and affluent urban consumers. The company aggregates the most engaging Web sites to serve display banners, rich media, video, and other online advertising. The most influential urban content Web sites, Web zines, newsletters, blogs, and forums belong to their network. What we don't understand is why more marketers don't belong to these and other press services. Yes, it's a fine thing to send your press releases to the general press for AAA-targeted events; you will likely reach a nice portion of your intended audience that way. But the primary reason to go out of your way—and to spend the few dollars a month it takes to engage such a service—is more compelling than any reason not to do it: credibility.

One of the points we made about the value of, for example, running advertisements in *Uptown*, *Black Enterprise*, and *Jones* magazines, BlackGivesBack.com, theroot.com, or becoming affiliated with a Black-oriented blog is that the readers appreciate that you are coming to their backyard to play. You're not expecting them to search you out; rather, you have sought them out to issue the invitation to use your brand. Moreover, the mere fact that your brand is among those included in these AAA-targeted media outlets carries a deeper meaning: that the brand has been vetted. *Uptown*, *Black Enterprise*, or *Essence* has done the evaluating for its readers and concluded that the brand intends their best interests.

The same holds true when news of your philanthropic donation, retail event, or conference sponsorship appears in the Black press. By virtue of the fact that the news is appearing in the ink or over the airwaves of a targeted newspaper or station confirms its worth within the community. Courting the Black press is a straightforward but often neglected step—a piece of your AAA marketing plan you can't afford to overlook. It serves as a viral component to your plan that reaches a far larger audience than the donation, event or sponsorship did and can extend quite far.

THE HUMAN FACTOR: THE IMPORTANCE OF EMPLOYEE TRAINING

Affluent African Americans are going shopping in ever larger numbers. Unfortunately, they are still seldom sought after the way that other customers with money might be. Worse, however, is that they are sometimes, often quite inadvertently, discouraged from shopping in upscale venues.

Remember Julia Roberts in Pretty Woman? After being ignored in a chic Beverly Hills boutique, her character returned the next day, having spent a lot of Richard Gere's money elsewhere. She asked the sales people who had looked down their noses at her, if they remembered her and reminded them that they had refused to sell to her. She pointed out that they probably worked on commission, and then said what they were probably already thinking: "Mistake. Big mistake. Huge!" As funny as that situation might have seemed to the average moviegoer, affluent African Americans in the audience recognized an all-too-familiar traditional retail experience. The experience is so pervasive that even someone as accomplished as Reginald Van Lee, senior vice president of Booz & Company, is not immune.

In an interview for Diversity Affluence's monthly eNewsletter, *The Royaltons Report*, Van Lee had this to say about his "invisible" demographic:

> "It's obvious to those of us who attend events, where the room is filled with 700 to 800 African American millionaires and billionaires. But it requires a slicing and dicing of the data to see that, one, this group is not as small as it's perceived to be, and two, it is growing. They have disposable income, and they are not going to

give it to someone who didn't take the time to learn how to approach them in the right way. You want to be one of those brands that does approach them in the right way and does get their money."

How do you set yourself up as a brand that approaches this target segment in the right way? Largely by making sure your approach does not encompass just the ads you run or the events you sponsor. That it trickles down to your consultants, your agencies, and every employee who is staffing your sales floor. When we asked Van Lee if he's ever had any bad experiences at the retail level because he was Black, and if so what he'd done about it, he was ready with an answer:

"Oh, yes! I've asked to see the manager, written letters, and sometimes quietly walked out but told everyone I knew about the incident. It's more in the past that I've lashed about and been upset. Now, I pretty much ignore passive racism, unless it's really offensive and aggressive. In most cases, I just leave and they lose my money."

Let us repeat that, so it will really stick in your mind: *"In most cases, I just leave and they lose my money."*

One participant in a Diversity Affluence focus group told the following story. She had decided to purchase a luxury vehicle, was pre-approved for a loan, and visited her local dealership ready to buy. She was treated so poorly by the sales staff that she left and made plans to travel over 800 miles to a dealership in her hometown instead. She was even willing to pay a premium price and additional shipping charges to be treated with respect and have a more pleasant experience. After she told her story, a full 90 percent of the focus group participants shared similar experiences of being treated poorly or indifferently at auto dealerships or in other retail environments. Remember: the brand gets you there, the customer service keeps you there.

The sad reality is that marketers often unwittingly fulfill an expectation that lurks in the mind of every African American: that they are not acknowledged or sought after or even plain wanted as customers. It is the modern equivalent of not being waited on at the lunch counter. Only now, it's the shoe counter at Saks Fifth Avenue or the makeup counter at Neiman Marcus. Most of the marketing insights we've given you in this book are external—things you can do outside of your corporate environment to attract consumers. But what you do *inside* the corporate environment is just as critical. Your brand must embrace diversity inside as well as out. By having someone at your own table who represents this target market—again, your consultant, your agency, your employees—you can better sustain the diversity efforts you've undertaken and keep them authentic. And you can avoid the frustrating fate of investing in a corporate marketing plan only to have it foiled by untrained employees at the retail level. We suggest to our clients that they follow a program similar to the innovative one that was instituted at the MGM Mirage, one of the world's largest hotel and gaming companies.

Debra J. Nelson is vice president of Corporate Diversity, Communications, and Community Affairs for MGM Mirage. She's responsible for the implementation of the company's diversity initiative and public relations, as well as governmental relations and corporate philanthropy. Known for her fearless efforts in creating diversity "pilot programs" during her career—which has included executive positions with Daimler Chrysler and Mercedes-Benz—we asked her to sit down and talk about the importance of diversity in corporate America, especially as it relates to a company's work force training.[34]

The MGM Mirage diversity initiative was launched at the end of 2000, as the chairman and CEO, Terrence Lanni, became aware of the changing demographics of both the hotel's marketplace and its employee population. In addition, research had revealed that the Mirage's customer profile included more multicultural consumers, who had higher education levels, and higher income levels, than the company had previously recognized. "These consumers have more and more dollars at their disposal and so have more choices to consider about where they play, stay, are entertained, and conduct their conventions and business meetings," Nelson explained. "By not embracing these changes, it affects success and profitability. Not only do we want to increase market share, but we want to create an environment where our customers feel welcome and want to come back and to share that experience with their friends." It was clear that a diversity initiative was a critical economic priority. "We established a board-level committee to lead this work, and we are still one of the few American companies that have such a committee. Following that, came a deliberate infrastructure to support the work, which included my position," Nelson said. And what was the work? "Diversity here [at the MGM Mirage] has three principle objectives: to position our company as an employer of choice, to position our properties as destinations of choice, and to secure a competitive position in the marketplace when it comes to doing business with minority, women and disadvantaged business enterprises."

One of the tools the Mirage makes use of toward this goal is their "diversity championing training," which Nelson is quick to cite as one of the company's best practices. It is three days of training "to help employees better understand the business case for diversity. We talk at the corporate level about why this work is important, but our employees are the first line of communications with our customers."

Your program doesn't have to be as formal as the one instituted by the MGM Mirage—your company, indeed, may not be as big or as complex and such a program wouldn't be practical. But the knowledge you, as a marketer, are gaining about the AAA market can't stay with only you. You'll certainly have to pass on statistics you are learning about the vast size of the segment, as well as your new insights in how to appeal to it, to your clients as you work with them to develop the pilot programs and marketing partnerships that will increase their customer base. But don't let the knowledge flow stop there. Pass it on to your client's employees. Make sure that when you help your client to bring all these new customers into his or her place of business, the first people who greet them at the threshold know how valuable their patronage is, and how to treat them as valued customers.

COMMON PITFALLS AT THE RETAIL LEVEL

To stress just how important the total retail experience is in successfully attracting affluent African Americans, let us tell you the three top negative retail level impressions that participants in Diversity Affluence's regional focus groups have reported—based on their all-too-common experiences during visits to upscale retail stores. These are the three main problems that marketers must be sensitive, cautious, and knowledgeable about in order to avoid alienating these potential customers.

1. Disrespect/racism. Whether perceived or real, disrespect and racism are quite tangible and all-too-familiar to most AAAs. One typical story involves entering a luxury retail situation dressed casually and being ignored by retail staff. To the contrary, the person relating the story had experienced a very positive reaction from sales staff when entering the same store, on a different day, but dressed more upscale or professionally. This experience was not always perceived as racist, as respondents were aware of friends who were not African American yet had similar uncomfortable experiences.

    Even when they were not ignored, many AAAs were treated in a way that was often disconcerting. Another focus group participant told a story about asking to see a "luxury brand" shirt in his size only to have the salesperson say, "That is expensive." This was a typical response noted by almost all focus group participants. Overt disrespect and racism is obvious; however, subtle disrespect or racism can just as effectively ruin the brand experience. It only takes one untrained or insensitive salesperson to undo corporate marketing efforts. Even more frustrating is that unless someone writes to corporate to complain, corporate will never know what has happened—and the overwhelming evidence suggests that AAAs won't write to corporate; they'll just tell one hundred of their closest friends. "When I shop at a luxury retail store dressed like a lawyer," said another participant, "I get good service. It's different when I go in casual clothing after dropping the kids off at soccer practice."

2. Aggressive Salespeople or Being Overlooked. Another common negative luxury-shopping experience is aggressive salespeople who seem more interested in making a sale than in meeting the prospective customer's needs. This kind of unwanted attention often made respondents feel uneasy because they were quite obviously being watched a bit too intently from a distance by staff. The insinuation was that they warranted scrutiny as potential shoplifters with obvious racist profiling overtones.

3. Insufficient Knowledge. It is understandably frustrating for participants who encountered sales staff at luxury retail stores who did not seem to know much about their products or services. This shortcoming was more expected and forgivable in lower-end retail outlets, but inexcusable at luxury retail shops—and AAAs have a particular sensitivity to this problem.

Of course, the MGM Mirage does more than make its corporate case for diversity to its employees. It actively seeks out and recruits emerging segments to its properties. Nelson cited a particularly relevant example of one of their outreach strategies, as it touches on many of the tactics we have been discussing in this chapter:

> "We did a pilot program at the National Urban League's national conference. We wanted to create a more direct return on investment, beyond putting our company name on the wall. So when we sponsored the Black Enterprise Women of Power luncheon, which was attended by approximately 1,000 people, we included a beautiful promotional piece in their gift bag, with a limited time offer on a few hotel packages. This cost us just under $2,000 for printing and graphics. Within the three-month window of return, we received about $50,000 worth of business."

Again, we stress the bottom line: $50,000 worth of business for an investment of just under $2,000. These are the kinds of numbers you too can achieve when you focus just a portion of your resources on the AAA segment. At the end of this book, you will find a practical list of do's and don'ts that will help you to approach this growing and coveted segment with confidence and competence. Our aim is to shorten your learning curve so that you will waste no time in engaging the AAA audience and achieving your own remarkable ROI.

Conclusion: Shortening the Learning Curve—Steps to Perfecting a Diversity Marketing Program

In February 2008 *New York Times* article, Stuart Elliot wrote, "The Interpublic venture, which is to be announced on Friday, is indicative of the intensifying interest on Madison Avenue in minority consumers."[1] When Madison Avenue acknowledges the merit and relevance of a trend, marketers and industry experts everywhere respond accordingly.

But the sum of all the parts is that although affluent African Americans wield $87.3 billion in buying power, they are still a relatively untapped target segment for any marketer, luxury or non-luxury. Whether you are Target or Ferragamo, Whole Foods or Vera Wang, you can benefit from acknowledging, understanding, and marketing to African Americans. How well you act on this opportunity will determine your success or failure. Brands that foster conversation among these passionate consumers will be rewarded with the word-of-mouth credibility. The bottom line is that it pays to market to affluent African Americans.

It doesn't matter if you have an entire AAA strategy, one pilot program, a tactic, or are simply trying to make your existing programs more inclusive. Just remember to test, tweak, refine, measure, repeat, and scale up, and scale out. Don't rest on your laurels and think you can pull the plug on the effort once it's successful. Remember that niche and target marketing is growing in importance. Don't confuse diversity hiring initiatives with diversity marketing. They are different, and deliver very different results.

From television programming executives to brand managers, business development professionals to media buyers, marketing executives, entrepreneurs and content providers, the time to act is now. Like any other consumer group, AAAs want marketers to tap into their lifestyle passions. Media partnerships, member-based organizations, and affinity groups will be the shortest distance from opportunity to results.

STEPS TO PERFECTING A DIVERSITY MARKETING PROGRAM: THE LOOK OF SUCCESS

While we've identified several brands poised for success among this segment—such as American Express, Audi, Apple, or Royal Caribbean—we settled on the one that resonated the most with AAAs: BMW.

The BMW brand is well-liked and well-respected by all luxury consumers—including AAAs. It's most definitely on their radar, and many AAAs are already customers. This is a brand poised to have an immediately successful opportunity with the AAA segment if it chooses to engage it. But when AAAs were asked about the brands they feel do a better job of marketing to them, BMW didn't make the cut. One focus group participant put it this way, "[As an African American,] you walk onto a BMW lot and you can wait all day [for someone to approach you]. But walk on a GMC lot and they swarm all over you." And no one we interviewed knew anything about the diversity programs BMW sponsored.

Table 6.1   African American Share of BMW

	2007 CYE	2008 CYE	2009 CYTD July 09
BMW New Vehicle Sales	259,376	220,361	92,981
African American BMW Sales	15,211	13,302	5,495
African American Share	5.86%	6.04%	5.91%

Source: Based on Personal Registrations Only—Does not include Fleet and Commerical Sales.

We saw this as an opportunity to paint a picture of what a BMW AAA initiative might look like. So, what is our dream campaign for BMW? How would we go about creating and executing a 360-degree campaign for this historic luxury brand?

CONDUCT A MARKETING AUDIT

The first thing we would do is conduct a marketing audit. A marketing audit conducted by an outside expert provides an organization with a "1000 feet in the air" perspective on their standing in the market among the competition with AAA consumers—and it identifies where they might already have assets and investments to leverage. It explores all facets of sales and marketing stakeholders—including event marketing (B2C and B2B), PR, CRM, Retailers, HR, Internet Marketing, Supplier Diversity, vendors, and agency partners who might be underutilized. It can help a brand to see if its company environment is too rigid and needs adjusting. It can also help a brand find money it didn't realize it had by looking into programs that aren't performing and repurpose those funds—or a portion of those funds—to pilot programs that reach new markets. In some cases, you will be creating relevancy where there is none, and that's the sign of a true marketing innovator.

Without the advantage of being able to conduct an actual audit of BMW or to sit down with its marketing executives and determine what their target audience is or could be, we have to do a little extrapolating from experience. Let's just say that an audit almost invariably brings to light four key insights:

1. An opportunity to leverage untapped niches likely exists.
2. You may already have assets and equities to expand on that you don't realize you have.
3. Funds from programs that are not performing or are underperforming can be repurposed to pilot programs.
4. You are likely leaving money on the table if you don't move swiftly.

An audit helps to assess exactly where the opportunities are, and where the risks might be, and how to mitigate them.

Taking our BMW example, let's just say the company is looking to attract AAAs who have an average, annual, per person income of $150,000. And, for the sake of our fantasy marketing campaign, let's unroll our pilot program in Atlanta, where we know the brand already has a presence and a local dealer willing to co-op the program. Let's also assume that we found $50,000 to $75,000 in a program that was underperforming and the company was willing to repurpose these funds for a pilot diversity program.

ESTABLISH PARTNERSHIPS

The first recommendation we would make to this venerable brand in its quest for AAA patronage, in terms of spending its funds wisely, is to form partnerships. These partnerships could be with philanthropic or cultural arts organizations that already enjoy the support of the local AAA community. Again we stress that in focus groups this segment repeatedly makes clear its preference to engage with affinity groups, member-based or charitable organizations on a local level. In Atlanta this might include the local chapter of the National Black MBA, but we would suggest that the brand reach out to an influencer for advice in outlining all the partnerships that are available to them. Remember that the goal is to develop a long-term relationship with an organization whose mission matches the integrity and DNA of the brand. It is often a prominent local figure who can best make wise recommendations and foster the personal connection between the organization and the brand.

We might suggest that BMW partner with a women's member-based organization because the AAA woman is always looking for a brand she can embrace. We might suggest an in-store retail event where the brand promotes a new product line, a ride and drive event, or a special pricing and lease offer incentive to the women of The Links or Mocha Moms.

ALERT THE MEDIA

But an organizational partnership isn't the only one we want to create. We also suggest that the brand partner with a local or regional Black-oriented media outlet. This could be a print venue, such as *Uptown*'s Atlanta edition; it could be an urban contemporary or smooth jazz radio station; it could be an online publication or blog such as The Black Socialite or www.blackgivesback.com. The point is to engage with local media that already has the eyes, ears, and *trust* of the local AAA community as both an advertiser and for potential editorial coverage.

Send pre- and post-event press releases to whatever media outlet your brand is partnered with, and don't forget to send them to the Black Press Wire and other luxury regional press. Make sure that a photographer, such as the well-known society photographer Patrick McMullen, knows about your event well in advance and has it on his or her schedule. They too can get the word out in post-event press.

HOST THE EVENT AND USE THE CONTENT

Within the partnerships that you establish, we would recommend creating a signature pilot event that could be tested in the brand's

Atlanta market and then be rolled out to other dealerships in cities such as New York, Washington, D.C., and Chicago. The event roll-out could be a cocktail reception, but we advise that in creating your first diversity event to remember that the goal is to create a template for an event that you can employ more than just once. Your aim is to expand it, so we often recommend starting with a concept that is eminently measurable and scaleable, one that resonates universally with this consumer across the country.

The BMW scenario is a perfect example of another advantage of using an event to create a retail experience: it gets the AAAs into your store. It lets your guests meet the company's sales staff and helps them build a relationship with the brand at the retail level—with, for example, BMW's financial services experts, top sales staff, and dealership owner. Business cards are exchanged, data is captured, and the beginnings of a personal relationship between the AAA and someone at the retail level, who she can call by name when she telephones, are established.

The in-store event must be very upscale, starting with an invitation, which, as we noted in chapter 5, should be handsomely packaged and hand addressed. The invitation list for this event could be culled by, for instance, combining an upscale publication's select subscriber list as well as the charity's or organization's, but the usual scale is to invite 150 to 200—or to invite 125 members of an organization and add a viral component by inviting each guest to bring a friend.

The women would be asked to give feedback to the marketing team, or other retail personnel. This is an important component of such a marketing event—how often do a brand's employees get firsthand feedback from consumers in such an amenable setting? A percentage of the money the women spend during the event will go to the charity and, within five business days, each guest receives a simple personal thank you note from the sales person with whom they connected at the event. Further follow-through can include using the content from the event to place photos on your Web site, include stories in company and consumer newsletters, and incorporate into post event press releases.

ESTABLISH A PRM METRIX

It's important to mention that the thank you notes are real paper cards, handwritten, sent via the U.S. Postal Service. We point this out because another facet of the event is to devise a method to capture the e-mail address of the attendees. These e-addresses are for a purpose other than thank you notes but remember: nothing beats the personal touch. Including a "calligrapher" line item in your event budget would be peanuts compared to the results you get in return.

The primary purpose of data capture is to create a prospect relationship management (PRM) system. In this case, how will the dealership follow up with all of the potential consumers who attended its event, beyond the handwritten thank you note? How will it keep track of all of those new hand-raisers and prospects? The two critical benefits of the PRM system are (1) continued engagement with the hand-raisers and prospects who have already been personally introduced to the brand, and (2) to tailor your follow-up and outreach to these new consumers by responding to what they tell you they want, rather than to what you think they want. The PRM system that any one entity creates can be as simple or as sophisticated as the client likes, and the follow-through can take place through several different avenues, including through e-mail and the Internet.

Perhaps, like many luxury brands, yours has been dabbling in the Internet or social media aspect of marketing, but you aren't sure how to apply it. This isn't unusual; the luxury industry as a whole is nervous about the Internet, e-mail outreach, and social networking because luxury brands are experiential. How does a marketer convey the experience of luxury on a flat computer screen? No one has quite figured out just yet what represents industry best practices.

Through audit of another client's marketing practices, however, we found a good practice that could be expanded to include the AAA segment. This brand had already started to do regular e-mailings to a growing customer list, and that was encouraging. We advised them to take that marketing tactic that they were already doing and apply it to their pilot diversity program and this audience. For BMW, we suggest a similar strategy: within the second week following the maiden event, each attendee would receive an e-mail promoting additional items from the line, or maybe announcing the debut of a new series depending on the timing of the event. This e-mail would be a specific call to action, with content tailored specifically to AAAs—possibly using the content of an existing e-mail promotion but tweaked to fit this audience. Clicks in this context are countable, a way to measure the response to this e-mail. Sales people should be alerted that feedback from these new customers must be collated as it comes in, as it represents valuable information that can be used in developing and growing the company's diversity program. This e-mail should also include a special code the buyer can apply when they purchase online; the code could offer the buyer a discount but, as critically, it would be able to be tracked, providing another method to measure the effectiveness of the e-mail promotion as well as the pilot program in general. Alternatively, BMW could sponsor a follow-up e-mail newsletter produced by the promotional partner where the reader can click through to a special offer.

As Christopher Vollmer, a partner in Booz & Company and the author of *Digital Darwinism*, has pointed out to us, luxury brands are recognizing that digital marketing is becoming more important to their businesses, but they are still learning how best to take advantage of it. More and more, luxury brands realize that their consumers are online and as a result they need to be there, too. However, the reality is that few luxury brands possess the necessary digital content and social-media skills to take full advantage of the

Web. Most brands are still experimenting and trying to figure out what works. As a luxury brand, adapting to digital now will put you ahead of the curve.

But however your company or client may choose to handle it, creating a prospect relationship management system to capture leads or hand-raisers from on- and offline events and establishing ways to get AAAs to a state of purchase readiness is critical. After which you can put them into a CRM system. Otherwise you'll have spent time, money, and resources with no leads to continue to engage beyond the "moment" of that event. That's unwise. Remember this audience likes personalized service and engagement—so stay engaged with them. It is a lost art of marketing best practices.

MEASURE, MEASURE, MEASURE

Calculating the impact of your pilot program on your company's bottom line is crucial in your ability to sustain and substantiate your diversity program. So decide well in advance of your event what yardstick you will use to determine whether or not your program is a success, how the events are measured up, and employ it rigorously. What constituted your minimum acceptable return on investment? If you invested, for example, $30,000 to create your pilot template, and your first event produced a return on investment of $85,000, would you consider your new diversity program one that warranted expansion? What percentage of e-mail respondents will you consider a success? On average, one can expect a 3 percent to 5 percent open rate from a third party, consumer-based e-mail campaign. If you could track a 10 percent to 15 percent response from targeted prospects, would that inspire you to continue the program in other retail venues in other cities? Quantify the company's investment and as well. Define key learnings from the pilot program with an eye toward tweaking, refining, and scaling the program up and out for the future.

TEST, TWEAK, REFINE, MEASURE, AND REPEAT (TTRMR)

Since budgets are always going to be an issue, we think it's important to create test markets or small pilot programs that give you a
measurable ROI. This way you can test the components of your custom diversity program and measure, tweak, and refine them so that
you can repeat the event with increasing success. As you become
comfortable with the inner workings of your program, and as your
confidence and results build, scale up and out—adding markets
and elements accordingly. Make sure you always have an established
way of measuring what you will accept as "success." Tweak, refine,
measure, and then do it again.

BE CONSISTENT

Once your pilot program has been refined and is a consistent success, continue to cultivate your new AAA audience with the next
level or layer of innovation. Do something different *again*. Your
next event may not necessarily be in-store. Maybe you'll host a networking event for a group affiliated with the Executive Leadership
Council. Maybe you'll sponsor a golf outing with the Council of
Urban Professionals or an awards dinner for Jack and Jill's top teen
volunteers and their parents. Or maybe you'll host a reception for
a fraternal organization. The idea of your second outing should be
focused around reciprocity: you have invited your audience to your
playground, and now you should give a positive RSVP when they
invite you to theirs. You should go to them for your second outing.

But whatever venue you choose, the key concept is consistency.
This does not mean that we believe every company should roll out
a series of similar events every month, or commit to purchase advertising in every issue of every publication we have recommended for
the roll out period—especially if there are budgetary concerns. It
does mean that once the company begins to engage with the AAA
segment, it continues to do so in some way, small or large, on a

regular, dependable basis. It means that it continues to speak to the people who support it with their purchasing power, and in proportion to the loyalty the emerging segment is demonstrating in return. Think creatively, and also think grassroots and frequency. If you approach this with an "I'm going to comply with our diversity quota" attitude you will only be hurting yourself. Why? This is an audience that brands and businesses can no longer afford not to engage. The money being left on the table can be snatched up in a nanosecond by your competition.

It reminds us of story of a client who, after an analysis of its business determined that one of its popular products held 53 percent market share among women. That was without any targeted effort. Even though a clear spotlight was shed on the subject, they still refused to target market and increase that 53 percent to a possible 75 percent. This would have blown the competition out of the water and helped them create an even stronger hold among a very important segment of the market—one that has proven time and again to be decision makers and to influence purchase decisions.

HIRE A CONSULTANT

The only way a program of the sort we are talking about will work is if your brand works with a seasoned 360-degree consultant to oversee all aspects of your diversity program implementation, measurement, and post event reporting. It's a simple fact—PR people think the world revolves around PR, and an ad agency or an event planner would come to the program with a narrower point of reference. The successful development and deployment of integrated marketing programs requires the skills and insights of someone who understands advertising *and* public relations *and* Internet marketing *and* social media *and* event planning—all of the degrees of an effective plan.

And, beyond providing a marketing audit that will help you to assess and leverage your brand's existing marketing assets, and apply them most efficiently to targeting a new market segment even after your first event roll out, this is the person who can help you measure the results of your program post-event and write the report you'll want to send to your executive management team to gain their support to try to duplicate your success in other markets, turning your pilot program into, essentially, a turnkey operation. This is the person who you can turn to to help you develop your business as you court your new AAA audience—and even as you expand your reach to include other Royalton groups. They will effectively help you avoid missteps, mistakes and misperceptions.

ESTABLISH A STANDARD OPERATING PROCEDURE

As you grow your business through expanding brand loyalty among AAA groups, even if it's on a very small scale, establish a standard operating procedure (SOP) for evaluating the opportunities that present themselves. How will you respond to them, engage in them, measure their success in realistic terms, and do so in a way that matches your corporate culture? Not every brand or business uses the same yardstick. For example, is it important for your brand to capture data? Should product specialists always be on site? What about a product display? How will the consumer interact with your brand? Are you driving traffic to a retail location or online? Does your business/brand offer specialized services such as a corporate giving or a personal shopper and is that communicated to your customer and prospect? Do you keep track of proposals that come your way and the opportunities they represent? How will you and upper management measure "success" and be realistic about what success really is at different stages of your involvement with an organization or event? The vibrant, responsive audience you are courting requires marketing that is just as vibrant and responsive.

The audit helps here, too. It will allow you to understand historic marketing decision-making patterns, where your efforts have been focused and if those are aligned with the current goals and objectives of your business.

BE NIMBLE

Create a slush fund in order to respond to last-minute opportunities. Most marketers (particularly luxury marketers) are quite rigid and don't realize the value of last-minute opportunities. Their complex corporate structures, guidelines, and policies cause them to lose out—and we are living in a world where three weeks to get an approval on the use of a log is absurd. Last-minute opportunities give you more leverage to negotiate price and elements. While you may lose the upfront promotional opportunity, you may get it on the back end with post-event press. It also provides the chance to create a category-exclusive scenario for your brand or business at a much lesser rate, thereby giving you the ability to become a true "marketing partner" the next year or event at a discount.

DON'T BECOME OPPRESSED BY RESEARCH

A lot of research today is flawed. That's just a fact. The research methods that are used are often out of date, and not much primary research on AAAs exists anyway. The research that does exist often does not ask the right questions, so the answers are incomplete or misleading. There are many multicultural researchers, experts, and consultants out there. Get to know them so they can help you understand the entire landscape of opportunity represented by this segment. Utilize all the tools at your disposal—polling, consumer intercepts, phone interviews with influencers, online surveys, and dynamic focus groups—to fully understand your AAA customer. Most importantly, use the research productively. We've seen many marketers order and analyze research—and then do nothing with what they've learned. They don't monetize the insight. Often this is because while they take the first step to find out about the AAA audience, they haven't hired a consultant who can actually introduce them and help their brand interact with this segment in a meaningful way. Zero effort means zero results. In the words of the famed comedian Flip Wilson, "You can't expect to hit the jackpot if you don't put a few nickels in the machine." It's possible to invest as little at $15,000 into research or consulting and translate that into $150,000 worth of business.

Don't let your analysis lead to paralysis. Sometimes you must simply listen to your intuition. Even the most data-immersed CMO still has to take a leap of faith at some point. After all, no amount of modeling can really predict how consumers will take to a new ad campaign or product. A consultant will help you merge the data and your intuition most effectively, most efficiently, and most accountably.

FROM MONUMENTAL TO INCREMENTAL

It doesn't require an advanced degree in calculus to see that the world's population growth, and accompanying demographic shifts, is exponential and that marketers will be affected by it. In fact, the new marketplace requires changes that are on a magnitude commensurate with those occurring in the population. Marketers must be cautious in not being superficial in their vision. As Drew Neisser, CEO of Renegade Marketing, put it, "This new kind of CMO is less interested in the monumental and more in the incremental...the CMO has evolved from a Chief Miracle Officer to a Chief Minutia Officer."[2]

Speaking of incremental, let's reflect for a moment on the 2008 Obama campaign. It raised a record-breaking $639 million, and much of it from small donors. It was an example of one of the most precise strategies and executions around today—the stuff Harvard

Business School best practice case studies are made of. They had one goal and many audiences—but they knew each audience intimately and this knowledge allowed them to reach out authentically to every one of them. You may be thinking "Yeah, but how much did he spend." That's not the point. The point is that he had a drive and desire to do it right the first time. And so should you, regardless of the size of your budget or initiative.

In this book we've pointed out the way toward reaching one underserved, very specific and affluent audience, and provided you with the insights you need to develop a plan to reach out in ways that resonate with them.

Are you ready to let marketing to AAAs pay off for everyone? All that's left for you to do is turn the last page, and go where the money is.

The Do's and Don'ts of Marketing to Affluent African Americans

Do: Spend time researching your target audiences' habits and motivators.

Don't: Assume that African American is a culture—it's a classification.

It's important to spend sufficient time researching your audience, understanding where they spend their time and how they communicate with each other. Do they read *Uptown* magazine and watch CNN? Are they interested in travel? Would they cross the street to go to Starbucks even if they are standing next to a Dunkin' Donuts? Giving in to stereotypes or assuming that you know what your audience aspires to based on old data is not a way to set yourself and your campaign up to succeed.

Do: Be bold and experiment with new mediums; you are much more likely to find your audience near the cutting edge.

Don't: Focus all your energy and budgets on any one tactical execution.

We live in a digital age, the key word being "we." It's not just one culture that is the entire population. While there are those that have more access to the technology, society as a whole has been redefined by innovations in communication (social media, mobile, etc).

We cannot afford to live in the past and expect our audiences to respond to antiquated marketing communications. To be successful in this day and age, we must be risk takers experimenting with new modes of communicating our messages to our target. The only risk you would be taking by "playing it safe" is the risk of destroying your brand.

Do: Leverage social media as there is no better place to find qualified, engaged African Americans.

Don't: Use social media once and then neglect the connections you make. Marketing today is not just about establishing communications with your audience, it's about maintaining a relationship with them.

Five hundred million social media users in the United States alone is a modest estimate, but it illustrates how relevant and fertile the social media landscape is. One of the toughest challenges is identifying your target within the digital spectrum. In social media, the audience tells the marketer (and everyone else for that matter) who they are and what they like. Wise marketers listen when their customers, and their potential customers, speak.

Do: Challenge mainstream, "general market" content distribution channels to find the "niche within." African American consumers don't just consume African American media and they resent the idea that people think they do.

Don't: Believe that marketing products to the general market captures African Americans. "Triggers" or motivators are radically different, and relying on broad communications to speak to a specific community is a huge waste of money.

As the way we communicate evolves, so does the way we consume our communications. The digital spectrum has opened up a massive world of content and hundreds of thousands (if not millions) of channels to retrieve that information from. The audience has become part of the contributors of content as well; blogging, social media, reviews, recommendations, and the like make it that much more difficult to pinpoint a consumer's habits. To assume that African Americans only look for African American content is ridiculous. However, a marketer must not lose sight of their target consumers' motivators. Marketing in the digital arena allows us to be highly efficient and tactical with our communications, therefore allowing us to speak to our audience where they spend their time—not just where we think they should spend their time. Ignorance is perhaps the biggest "don't." Without the proper research and understanding, many marketing plans are destined to fail even before they are launched.

Do: Have a multi-tiered, cross-platform media mix. Affluent African Americans are moving targets; if you are able to continue the "conversation" from one medium to the next you will have a higher return on your investment.

Don't: Be invasive. Understand and respect your consumers' boundaries. Allow your customers to tell you how far they will allow you into their personal space.

Given the vast array of communication mediums, a savvy marketer must utilize multiple mediums in order to effectively reach their target. There is a line that cannot be crossed however; we don't like to be encroached upon. The Internet has made everyone an explorer; we have the ability to search for answers, products, social connections and more. It is important to allow your audience to "discover" you; in this case they will be more likely to share their discovery with their friends. But bombard them with your messaging and they are more likely to warn their friends away.

Do: Use cultural entertainment, movies, and the like to integrate your branding message.

Don't: Rely on the same old entertainers, movies, and other stereotypical cultural touchstones. Expand your thinking and move beyond stereotypes and surface opinions. Being myopic or small-minded will yield small results. There are African Americans who like country music, tennis, and Nascar. Pigeonholing yourself by using a finite set of reference points, entertainers, and venues to speak to your target audience limits the success of your campaign. It all goes back to knowing what your audience wants and not being afraid to listen to them.

Do: Look for strategic partnerships that benefit both parties while striving for a common goal.

Don't: Try and do it alone. The media landscape is far too vast for anyone to get their arms around. A life raft in the middle of the ocean has little chance of reaching shore. While you will always want to be the head chef in your own kitchen, you must always remember that someone has to grow the food that you cook with. There are very few silos in the media world these days; don't allow personal pride to block your route to success. Look for companies or individuals that possess talents and skills that will improve your business and products, and find ways to utilize their talents and skills to further your success. Think of it like this: you're the best apple pie maker in the land but in order to make the best apple pies you must have the best apples—so you look for the best apple grower in the land and work together to make the best apple pies.

Do   Always, always, *always* push the envelope when it comes to finding new ways to build a relationship with your audience.

Don't: Play it safe. If you're not proactively working to achieve your goals, know that your competitors are.

Technology and the way we as a society utilize technology is always changing. We cannot afford to wait for society to tell us when they are ready because the only thing that society is ready for is change. Looking for ways to connect with your audience is not unlike searching for new sources of oil, gold, or diamonds. We must always strive to find new ways to reach our audience—ways that will be welcomed and well received. Like anything else, if you're not doing it, someone else almost certainly is.

Do: Look for ways to integrate and weave your brand into your audience's lifestyle. Allow yourself to be "available" when your consumer is most receptive.

Don't: Force your message. The majority of consumers are not inclined to think well of a product or service if they feel commanded to do so. Instead, develop a rapport that allows the consumer to express himself and in return you will get all the information you need to develop a customer for life.

The Internet has made marketing a 24/7, 365-day-a-year opportunity. If you're not "there" for your customer when they want you, you have missed an opportunity to further develop trust and acceptance. Technology allows us to stay connected even when we sleep. Ignore the opportunities to be available for your audience and your audience will most definitely ignore you.

Do: Work hard to convert a consumer's "wants" into "needs."

Don't: Confuse "wants" and "needs" as the same thing.

The economic temperature of the world we live in, coupled with the rapid evolution of technology in communication, has created a new challenge for marketers. While the landscape and modes of communications have changed, basic marketing principles are as relevant now as they were a hundred years ago. It is up to marketers to understand the tools that are available to us and to utilize those tools to create our successes. Imagine what Galileo or Mozart would be able to accomplish today with the tools that would be available to them.

NOTES

INTRODUCTION: WHERE THE MONEY IS

1. *Where the Money Was: The Memoirs of a Bank Robber*, Viking Press, 1976
2. Christopher Vollmer, "Digital Darwinism," strategy-business.com, May 2009, Viking Press, 1976
3. Ibid
4. "The Gift Economist," Deborah Solomon, *The New York Times Magazine*, July 19, 2009
5. National Association of Home Builders; www.nahb.org
6. Melanie Shreffler, Research Alert Special Report, "Snapshot: Black, Hispanic and Asian Americans," March 2009
7. Greg McBoat, Chief Economist, Diversity Affluence, "Affluent Ethnic Consumers," December 2008
8. "The End of White America?", Hua Hsu, *The Atlantic*, January/February 2009
9. "Race in 2028," Ross Douthat, *The New York Times*, July 20, 2009
10. Bill Clinton, address to students at Portland State University, 1998
11. Ibid
12. Ibid
13. "African-American Market in the U.S.," Packaged Facts, February 2008
14. "The End of White America?", Hua Hsu, *The Atlantic*, January/February 2009

15. Ibid

16. "Guess What, America? There is a Black Middle Class," Moses Foster, *Advertising Age*, May 19, 2008

1   CORPORATE AWARENESS

1. http://www.smith-winchester.com/branding.html

2. http://www.marketingprofs.com/5/nicastro4.asp

3. *Purple Cow: Transform Your Business by Being Remarkable*, Seth Godin, Portfolio, 2003

4. "The End of White America?", Hua Hsu, *The Atlantic*, January/February 2009

5. http://www.sassybella.com/2007/12/pharrell-williams-for-louis-vuitton-jewelry-to-be-finally-released-in-2008/

6. http://www.blackgivesback.com/2009/09/mary-j-blige-and-gucci-partner-for.html

7. "The Substance of Style," economist.com, September 17, 2009

8. *Journal of Advertising*, "Ethnic Evaluations of Advertising: Interaction Effects of Strength of Ethnic Identification, Media Placement, and Degree of Racial Composition," Corliss L. Gren, March 22, 1999

9. Selig Center for Multicultural Economy Report, "African-American/Black Market Profile," 2007

10. Association of National Advertisers, survey, November 13, 2008

2   THREE DEGREES OF SEPARATION: EVERYONE KNOWS EVERYONE

1. The *New York Times Sunday Book Review*, "Visible Young Man," Toure, May 1, 2009

2. Levi Fox, Gretchen Sund, Caroline Altman, "Urban and Urbane: The *New Yorker* Magazine in the 1930s," American Studies Program at the University of Virginia

3. http://www.nps.gov/archive/elro/glossary/great-depression.htm

4. "Who Needs the NAACP?", Benjamin Sarlin, *The Daily Beast*, July 10, 2009

5. Ibid

6. Ibid

7. Ibid

8. The Museum of Broadcast Communications, Mary Ann Watson

9. *Revolution Televised: Prime Time and the Struggle for Black Power*, Christine Acham, University of Minnesota Press, 2004

10. *Provocateur: Images of Women and Minorities in Advertising*, Anthony Joseph Paul Cortese, Rowman & Littlefield, 1999

11. "Alcohol and Cigarette Advertising on Billboards," D.G. Altman, C. Schooler, M.D. Basil, Stanford Center for Research in Disease Prevention, Stanford University School of Medicine, Health Education Research, Vol. 6, No. 4, 487–480, Oxford University Press, 1991

12. "Young Blacks Disproportionately Exposed to Alcohol Ads, Study Says—Noteworthy News," *Black Issues in Higher Education*, July 17, 2003

13. Ibid

14. "New Book First to Tell Inside Story of Blacks in the Advertising Business," Craig Chamberlain, Newsbureau, University of Illinois at Urbana-Champaign, February 11, 2008

15. Ibid
16. http://national.jackandjillonline.org
17. Ibid
18. *Our Kind of People: Inside America's Black Upper Class*, Lawrence Otis Graham, Harper Perennial, 1999
19. "Meet the New Elite, Not Like the Old," Helene Cooper, *The New York Times*, July 26, 2009
20. Ibid
21. "Minority Enrollment in College Still Lagging," Mary Beth Marklein, *USA Today*, October 30, 2006
22. *Survival of the Black Family*, K. Sue Jewell, Praeger Publishers, 1988
23. "Black in America," Soledad O'Brien, CNN, July 2009
24. "Wealth in Black America," Northern Trust, July 2008
25. http://research.lawyers.com/Estate-Planning-Survey.html
26. *Why Should White Guys Have All the Fun? How Reginald Lewis Created a Billion-Dollar Business Empire*, Reginald Lewis, John Wiley & Sons, 1994
27. "Meet the New Elite, Not Like the Old," Helene Cooper, *The New York Times*, July 26, 2009
28. "Black Wealth/White Wealth: An Issue for the South." Scott Doron, Elaine Rideout Fisher, Economic Investment Strategies (EIS), Associates, 2003

3 MEET THE ROYALTONS

1. Phoenix Cultural Access Group, 2008 U.S. Diversity Markets Report
2. Yankelovich, MONITOR Multicultural Marketing Study, 2009
3. http://online.wsj.com/article/SB124052234831749871.html

4 INSIGHTS, TRENDS, AND 360-DEGREE MARKETING

1. Robert McIntosh, Canadian Marketing Blog, Canadian Marketing Association, October 15, 2007
2. Brand Strategy, Will Harris, Nokia marketing director for the United Kingdom, October 13, 2008
3. "Is 360-degree Marketing Dead?," Ruth Mortimer, brandstrategy
4. "Burberry to Launch Social Networking Site," Jennifer Creevy, *RetailWeek*, September 17, 2009
5. "Draftcb NY Combines Media, Digital and CRM to Launch Real-Time Marketing Group," Draftcb press release, June 9, 2009
6. Unity Marketing, *Luxury Business Newsletter*, October 1, 2009

5 PUTTING INSIGHT INTO ACTION

1. "Unity Marketing's Latest Survey of Affluent Consumers Points to Signs That the Luxury Consumers are Beginning to Recover From the Recession," *Marketwire*, July 24, 2009
2. Lucius T. Outlaw, *Notre Dame Philosophical Reviews*, June 2009, review of *The Browning of America and the Evasion of Social Justice*, Ronald R. Sundstrom, The Browning of America and the Evasion of Social Justice, SUNY Press, 2008
3. "Black American Personal Wealth," Consumer Federation of America, 2002
4. Black Wealth, White Wealth, Melvin Oliver, 2/E, Routledge, 2006

5. "Warmed-over Myths of Black Wealth," Alfred Edmond, Jr., *Black Enterprise*, February 27, 2009

6. Ibid

7. *The Affluent Consumer: Marketing and Selling the Luxury Lifestyle*, Ronald D. Michman and Edward M Mazze, Praeger Publishers, 2006

8. "Advertising Targeting African-American Consumers Exceeds $2.3 Billion Annually," emergingminds.com, January 28, 2008

9. "Radio Snares Biggest Share of African American Ad Dollars," mediabuyerplanner.com, January 29, 2008

10. Press release; http://www.aetna.com/news/newsReleases/2009/0720_AtlantaCity.html

11. http://www.mochamoms.org/article.html?aid=9

12. US Census Bureau, 2005 estimates; processed by the Atlanta Regional Commission (ARC)

13. "Wealthy Blacks are Latest Target for Nations Bank," Kelly Greene, *Atlanta Business Chronicle*, October 4, 1996

14. Ibid

15. Ibid

16. Ibid

17. *TED* (Technology, Entertainment, Design) is an invitation-only event where the world's leading thinkers and doers gather to find inspiration. www.ted.com

18. "Q&A: Why Lexus Is Spreading Sustainability Message to African Americans," Becky Ebenkamp, Brandweek.com, March 10, 2009

19. Ibid

20. Ibid

21. "Lexus Targets Affluent Black Women," Lauren Bell, DMNews, July 27, 2009

22. "Victoria's Secret Pink Adds HBCU Flair," BlackEnterprise. com; Posted in Business News on July 24, 2008
23. Ibid
24. "Better Late Than Never—Royal Caribbean Cruise Line to Target Advertising Toward African Americans Through Chisholm-Mingo Group," Nicole Marie Richardson, *Black Enterprise*, October 2000
25. Ibid
26. The True Agency, Infiniti case study
27. Ibid
28. Ibid
29. Ibid
30. Ibid
31. "Campbell's YSL Tribute," Vogue.com, June 3, 2008
32. "Yves St. Laurent, Givenchy and the Black Models of Today," Angela L. Serrette, sojones.com, June 11, 2009
33. http://www.time.com/time/magazine/article/0,9171, 1005550–3,00.html
34. 2008 interview from the Royaltons Report

Conclusion: Shortening the Learning Curve—Steps to Perfecting a Diversity Marketing Program

1. http://www.nytimes.com/2008/02/08/business/media/ 08adco.html
2. http://www.diversitybusiness.com/news/diversity.magazine/ 99200841.asp

INDEX